Lecture Notes in Computer Science 16951

The series Lecture Notes in Computer Science (LNCS), including its subseries Lecture Notes in Artificial Intelligence (LNAI) and Lecture Notes in Bioinformatics (LNBI), has established itself as a medium for the publication of new developments in computer science and information technology research, teaching, and education.

LNCS enjoys close cooperation with the computer science R & D community, the series counts many renowned academics among its volume editors and paper authors, and collaborates with prestigious societies. Its mission is to serve this international community by providing an invaluable service, mainly focused on the publication of conference and workshop proceedings and postproceedings. LNCS commenced publication in 1973.

Maarten Löffler · Alice Raffaele ·
Francesco Silvestri
Editors

Creative Mathematical Sciences Communication

8th International Conference, CMSC 2026
Padua, Italy, October 12–15, 2026
Proceedings

Springer

Editors
Maarten Löffler
Utrecht University
Utrecht, The Netherlands

Alice Raffaele
University of Padua
Padua, Italy

Francesco Silvestri
University of Padua
Padua, Italy

ISSN 0302-9743 ISSN 1611-3349 (electronic)
Lecture Notes in Computer Science
ISBN 978-3-032-38886-5 ISBN 978-3-032-38887-2 (eBook)
https://doi.org/10.1007/978-3-032-38887-2

This Springer imprint is published by the registered company Springer Nature Switzerland AG
The registered company address is: Gewerbestrasse 11, 6330 Cham, Switzerland

Preface

This volume contains the papers presented at the 8th Edition of the Creative Mathematical Sciences Communication (CMSC 2026) conference. The conference was held at the University of Padua (Padua, Italy) from October 12th to October 15th, 2026. This conference marked another milestone in our ongoing journey to enhance computational thinking through innovative, constructionist, and hands-on activities. Since its founding in 2013, CMSC has been a platform for fostering creativity and collaboration in the realm of computer science and mathematics communication.

Historical Context

The CMSC conference has a rich history, with previous editions held in Trier (Germany, 2024), Bergen (Norway, 2022), Poznań (Poland, 2020), Wellington (New Zealand, 2018), Lübeck (Germany, 2016), Chennai (India, 2014), commencing in Darwin (Australia, 2013), initiated by Michael R. Fellows and Frances Rosamond. Over the years, the conference has taken on a biennial rhythm. Each of these gatherings has contributed to the growth and evolution of our community, expanding our understanding of and approach to teaching computational thinking. This initiative can be traced back to the advent of "Computer Science Unplugged".

Conference Mission

The CMSC conference aims to explore new creative ways to communicate the rich mathematical ideas that underpin computer science, whose foundations are, at their core, mathematical. Mathematicians and computer scientists share a growing need and interest in outreach and engagement at all levels. By bringing together researchers, educators, and communicators, the conference seeks to explore new approaches to mathematical-computational thinking and lay the groundwork for innovative methods of teaching and research engagement across all levels of education. The organizers of CMSC believe that engagement in the mathematical sciences runs in two directions. Certain areas of research, such as algorithms and combinatorics, are accessible even to elementary-age children, while conversely, interaction with non-specialists, including young learners, can spark new research questions. Creating opportunities for broad and inclusive engagement in the mathematical sciences not only enriches researchers but also brings lasting benefits to society as a whole.

The topics of the papers presented at CMSC 2026 are diverse and include, for instance, unplugged approaches to computer science concepts, creativity for computational thinking, teaching artificial intelligence, game-based learning, technology-enhanced tools, and developing higher-order reasoning.

Initiatives in CMSC 2026

CMSC 2026 continued the initiatives started at CMSC 2024, namely peer-reviewed research papers published in the LNCS proceedings, keynote speakers, and a one-day workshop called *Teachers' Day* to engage the local educational community, especially primary and secondary school teachers. For CMSC 2026, we also added a special event on generative AI and co-creativity in music.

Peer-Reviewed Research Papers

An opportunity for authors at CMSC is to publish their research papers, experience reports, and activity descriptions under a blind review process. This initiative ensures that the insights and innovations shared during the conference are preserved and accessible to future generations. Authors were invited to submit manuscripts on topics related to the conference's scope. More precisely, CMSC welcomed the following contributions:

- Standard papers from 4 to 12 pages in one of the following categories: research papers (e.g., empirical, qualitative, quantitative, theory building, research methods, comparative studies, transferability of methods and results from other disciplines); best practice papers/country reports/experience reports; description of creative products, for instance, dance or song performance; position/discussion papers.
- Poster proposals up to 3 pages for work in progress, short description of published work.

Contributions were peer-reviewed by members of the Program Committee. For standard papers, we took a 2-round approach, with two different deadlines. Papers submitted in the first round not only received a binary accept-reject decision, but some authors were invited to revise their manuscripts according to the reviewer comments and then resubmit the paper to be reviewed a second time. In the first round, we received 10 submissions: 3 papers were accepted for publication in the proceedings, and 2 were asked to resubmit a revised version in the second round. In the second round, we received 25 submissions, and 17 were accepted for publication in the proceedings. Each submission was reviewed in a single-blind process by 3 Program Committee members for quality, originality, creativity, and relevance. The authors of accepted papers were invited to present their works during the conference.

Poster proposals were handled differently: they were not included in the proceedings. The goal of posters is to present ongoing or previously published work, aiming to stimulate further discussion. The proposals were reviewed by the Program Committee, and the authors of accepted proposals presented printed posters during a session at CMSC 2026.

Keynote Speakers and Special Event

We are very grateful that we could attract three prominent researchers in the field as keynote speakers: Timothy C. Bell, Vanesa Guerrero Lozano, and Neeldhara Misra. The titles and abstracts of their talks are provided on the following pages.

Furthermore, CMSC 2026 had a special event on *Generative AI and co-creativity in music* by Sergio Canazza, Alessandro Fiordelmondo, Elisa Gremmo, and Ardavan Vossoughi (*Centro di Sonologia Computazionale*, University of Padua). The event brought together a small group of experts from the hard sciences (computer science and engineering) and the humanities (philosophy and musicology) to address these questions: Can machines compose? How does musical writing change when a composer works side by side with an AI?

Teachers' Day

The final day of the conference, Thursday, October 15, was dedicated to a hybrid Teachers' Day workshop, conceived as a space for a two-way exchange between the researchers attending CMSC 2026 and teachers from primary and secondary schools. The day reflected the conference's broader philosophy that engagement with mathematics and computer science flows in both directions: while contact with practicing teachers and their students can spark new research questions, researchers in turn bring accessible, hands-on approaches to topics such as algorithms and computational thinking into the classroom. More precisely, the program featured invited presentations, tutorials, and interactive workshops of interest to teachers. These activities also included some scientific articles accepted for the conference, allowing authors to present their work and illustrate how it can be used or adapted for a classroom activity. Activities were delivered in Italian or English. We were honored to welcome Centro Montessori SPES, Maria Francesca Carfora, Lorella Carimali, and Frances Rosamond as invited speakers.

Final Words

We extend our heartfelt gratitude to all authors for their valuable contributions and to the Program and Steering Committees for their diligent efforts and support. Their dedication was instrumental in ensuring the success of CMSC 2026. We hope that this volume, with collected conference proceedings, will serve as a lasting record of the many invaluable discussions and continue to inspire future generations to remain creative in communicating about the mathematical sciences.

August 2026

Maarten Löffler
Alice Raffaele
Francesco Silvestri

Organization

Program and Organizing Chairs

Maarten Löffler	Utrecht University, the Netherlands
Alice Raffaele	University of Padua, Italy
Francesco Silvestri	University of Padua, Italy

Steering Committee

Frances Rosamond (Chair)	University of Bergen, Norway
Małgorzata Bednarska-Bzdęga	Adam Mickiewicz University Poland
Timothy C. Bell	University of Canterbury, New Zealand
Sarah Carruthers	Vancouver Island University, Canada
Valentina Dagiene	Vilnius University, Lithuania
Michael Fellows	University of Bergen, Norway
Valia Mitsou	IRIF, Université Paris Cité, France
Jam Ramanujam	Institute of Mathematical Sciences, India
Rüdiger Reischuk	University of Lübeck, Germany
Eljakim Schrijvers	Eljakim IT, the Netherlands
Matt Skoss	Northern Territory Dept. of Education, Australia
Verena Specht-Ronique	HLTM Marburg, Germany
Brett Stephenson	University of Tasmania, Australia

Program Committee

Tim Bell	University of Canterbury, New Zealand
Sergio Canazza	University of Padua, Italy
Sarah Carruthers	Vancouver Island University, Canada
Valentina Dagiene	Vilnius University, Lithuania
Henning Fernau	University of Trier, Germany
Bernd Gärtner	ETH Zurich, Switzerland
Henry Hickman	University of New South Wales, Australia
Thore Husfeldt	IT University of Copenhagen, Denmark
Jyothi Krishnan	IT Gandhinagar, India
Martina Landman	University of Cambridge, UK
Lukas Lehner	TU Wien, Austria

Michael Lodi	University of Bologna, Italy
Zuzana Masárová	ISTA/Grow Trails, Austria
Monica Mattei	International School of Turin, Italy
Valia Mitsou	Université Paris Cité, France
Matthias Mnich	Hamburg University of Technology, Germany
Elena Mumford	Eindhoven, the Netherlands
Rüdiger Reischuk	Universität zu Lübeck, Germany
Giovanni Righini	University of Milan, Italy
Inge Schwank	University of Cologne, Germany
Jacqueline Staub	University of Trier, Germany
Jay Thakkar	IIT Gandhinagar, India
Josef Tkadlec	Charles University, Czechia
Sebastian Wild	Philipps-Universität Marburg, Germany

Additional Reviewers

Constantin Economides
Sonja Klein
David Sipos
Karolina Tammemaa
Mahmut Yavuz

Local Organizing Members

Mariafiore Tognon	University of Padua, Italy
Riccardo Zuech	University of Padua, Italy

Supporting Partners

We are deeply grateful to our sponsors and supporting partners, whose contributions made this event possible and accessible to a wider audience.

- Department of Information Engineering, University of Padua
- Bertinoro international Center for informatics – BiCi
- Associazione Italiana di Ricerca Operativa – AIRO
- Jane Street
- MaddMaths! ETS
- Motion Analytica
- Istituto Nazionale di Alta Matematica "Francesco Severi" – Gruppo Nazionale per il Calcolo Scientifico
- Unione Matematica Italiana
- Springer
- Yamaha

- Laboratorio Nazionale CINI "Informatica e Scuola"

CMSC 2026 was developed in collaboration with *Mathematics Without Borders*, an international education equity initiative sponsored by the Rotary Club of Taree, District 9660, and designed to improve access to computational thinking and STEM education for underrepresented groups, particularly women, early-career researchers, and participants from developing countries.

Invited Talks

Reflections on 30+ Years of Computer Science Unplugged

Timothy C. Bell

University of Canterbury, New Zealand

Abstract. Computer Science (CS) Unplugged emerged from independent efforts by Mike Fellows and Tim Bell in the late 1980s. Following their first meeting in 1992, these ideas developed into CS Unplugged, which was launched around 1996. Thirty years later, the unplugged approach is used internationally, discussed in hundreds of research papers, and embedded in school curricula around the world. Remarkably, many of the original activities developed in the 1990s remain as relevant today as ever. Although educators have long recognised the power of unplugged activities to engage learners, understanding why they work has been a longer journey. Research over the past three decades has shed light not only on the strengths of the unplugged approach, but also on its limitations and on how it can be used most effectively. In this talk, I reflect on the history of CS Unplugged, the debates and controversies that have shaped its development, and what we now know about the learning processes that occur when wide-eyed students encounter computer science through unplugged activities.

Mathematics Beyond the Classroom: Two Experiences in Science Communication

Vanesa Guerrero Lozano

Universidad Carlos III de Madrid, Spain

Abstract. Creative science communication can play a key role in making mathematics more accessible, relevant, and socially meaningful for young people. This talk presents two outreach initiatives developed in Spain that use different approaches to foster scientific curiosity, critical thinking, and awareness of the opportunities offered by higher education. The first project, Data Science Gymkhana, is an annual activity launched in 2020 within the STEM4Girls program at Universidad Carlos III de Madrid. Designed for secondary school students, the Gymkhana promotes data literacy and the scientific method through collaborative problem-solving. Each edition focuses on one of the United Nations Sustainable Development Goals, combining a keynote talk by a researcher with hands-on data analysis challenges that encourage participants to explore how mathematical and statistical tools can contribute to addressing real-world societal issues. The second initiative, Your Face Sounds Familiar, is an outreach program developed by researchers from Universidad de Sevilla and Universidad Carlos III de Madrid to bring science and the university experience closer to students living in rural areas. Through interactive sessions led by researchers from diverse disciplines, including mathematics, the project introduces participants to academic careers, showcases the role of research in solving contemporary problems, and seeks to reduce geographical and social barriers that may discourage young people from pursuing higher education. Although they target different contexts, both projects share a common goal: to present mathematics not as an abstract discipline but as a powerful tool for understanding the world, making informed decisions, and creating opportunities. The talk will describe the design of these initiatives, discuss the lessons learned from their implementation, and reflect on how creative dissemination activities can broaden participation in the mathematical sciences and inspire the next generation of students.

Building Joyful Interactives

Neeldhara Misra

Indian Institute of Technology, Gandhinagar, India

Abstract. Reading about an idea and wrestling with it are different cognitive acts. A well-designed interactive experience, either tactile or virtual, can potentially merge these aspects: when a learner plays around with concrete examples and experiments with parameters, the ideas involved become conjectures to test, fix, and potentially generalize. A productive interactive experience is not only about ornamental animation or frictionless learning; it is about agency, immediate feedback, and the productive surprise of being wrong in an informative way. Over several years of building small, purpose-built web interactives, I have found that key decisions are structural: what can the learner change, what remains fixed, and how quickly does the system reflect their move? The process of making interactives is iterative and AI-assisted coding has changed the economics: a pedagogical hunch can become an afternoon prototype, making experimentation cheap enough to take seriously. The human in the loop remains responsible for mathematical correctness, purpose, taste, and care. This talk will be a case study of interactive experiences (also known as explorable explanations) I have enjoyed working with, along with commentary on what makes them fun and useful.

Contents

Teaching AI

Game-Based Learning in Mathematics and CS

Technology-Enhanced Tools

Developing Higher-Order Reasoning

Unplugged Approaches to CS Concepts

A Practical Activity on the Shortest Path Problem with Theoretical and Algorithmic Implications

Pietro Belotti[1], Francesca Fumero[2], and Federico Malucelli[1](✉)

[1] Department of Electronics, Information and Bioengineering, Politecnico di Milano, Milan, Italy
{pietro.belotti,federico.malucelli}@polimi.it

[2] Department of Management, Economics and Industrial Engineering, Politecnico di Milano, Milan, Italy
francesca.fumero@polimi.it

Abstract. The shortest path problem and Dijkstra's algorithm constitute foundational topics in many basic courses of Operations Research and Computer Science. Yet, they remain conceptually challenging for many students: experience and literature document recurrent misconceptions, persistent difficulties in grasping algorithmic logic and optimality conditions and a limited understanding of the dual perspective. Learners often struggle to connect procedural steps with underlying graph-theoretic structures and to interpret shortest path problems within a broader optimization framework.

This paper proposes a structured manual activity designed for university-level courses to support conceptual understanding through direct physical engagement. Using inexpensive and tangible materials, students construct and iteratively update weighted networks, physically enacting the step-by-step execution of Dijkstra's algorithm. The activity is explicitly designed to make visible the role of tentative distances, the maintenance of feasibility, and the relationship between primal shortest paths and dual node potentials. Grounded in the active learning paradigm, the approach leverages well-known cognition principles according to which coordinated hand-brain activity can strengthen intuition and facilitate abstraction. By externalizing algorithmic processes into manipulable artifacts, students are encouraged to reason about optimality and dual feasibility in a concrete manner before formalization.

Keywords: Shortest paths · Dijkstra's algorithm · Active learning · LP Duality

P. Belotti—On the night of August 21, Pietro passed away after a brief battle with an incurable illness. We will always carry with us his scientific rigor, his humility, and his quiet generosity.

M. Löffler et al. (Eds.): CMSC 2026, LNCS 16951, pp. 3–15, 2027.
https://doi.org/10.1007/978-3-032-38887-2_1

1 Introduction

Graph theory and combinatorial optimization algorithms are central components of introductory courses in Operations Research and Computer Science. Despite their foundational role, these topics are often experienced by students as conceptually demanding. The main difficulty does not typically concern the mechanical execution of algorithmic steps, but rather the comprehension of the structural logic that guarantees correctness. As a result, students may learn to reproduce procedures without fully internalizing the reasoning principles that justify them.

The shortest path problem and Dijkstra's algorithm provide a paradigmatic example of this phenomenon. Empirical studies in computing education have shown that students often misinterpret the meaning of tentative distance labels, the state of permanently selected nodes, and the purpose of relaxation operations [11]. These misunderstandings can prevent learners from building a coherent mental model of the algorithm's behavior. Within an OR curriculum, an additional conceptual hurdle arises when students are asked to relate shortest path algorithms to the broader framework of optimization theory and, in particular, to duality. Although duality is usually introduced in the context of linear programming, the shortest path problem admits a natural primal-dual interpretation in terms of arc selection and node potentials. However, when duality is later presented in a purely algebraic form, students often fail to connect it with earlier algorithmic experiences.

These difficulties motivate the exploration of instructional approaches that strengthen conceptual understanding by making structural properties more explicit and observable. A substantial body of research in STEM education indicates that active learning strategies-characterized by structured student engagement in reasoning, discussion, and problem solving-consistently lead to improved academic performance and deeper conceptual understanding compared to exclusively lecture-based instruction [3,10]. Large-scale meta-analyses report higher examination scores and reduced failure rates in active learning environments across science and engineering disciplines. In addition to cognitive engagement, pedagogical traditions emphasizing manipulation and concrete exploration suggest that coordinated hand-brain activity can facilitate abstraction and intuition development [8]. In domains involving structural reasoning, such as graph theory and optimization, externalizing abstract constructs into tangible representations may help students bridge the gap between procedure and theory.

Motivated by these considerations, we propose a structured hands-on classroom activity aimed at supporting conceptual understanding of the shortest path problem and its dual interpretation. The activity relies on the construction of a physical weighted graph using inexpensive materials. Through guided manipulation of this model, students first identify shortest paths by direct experimentation. This exploration naturally leads to the formulation of a dual perspective in terms of node potentials and feasibility constraints, without initially invoking formal dualization rules. In a subsequent phase, the same physical representation is used to mirror the iterative logic of Dijkstra's algorithm, establishing a clear correspondence between physical node positions and tentative distance labels.

The contribution of this work is twofold. First, we present a concrete pedagogical tool that renders key algorithmic steps and optimality conditions visible. Second, we show how this experience can scaffold students' readiness for duality theory, supporting a smoother transition from intuitive reasoning to algebraic formalization in later stages of the course. Moreover, the activity is intentionally designed to be low-cost, easily reproducible, and scalable: it can be implemented even in large classroom settings while ensuring that each student actively participates in the construction, manipulation, and interpretation of the model.

2 Dijkstra's Algorithm: Challenges and Active Teaching Approaches

Understanding graph algorithms poses recurrent learning challenges for undergraduate students. Our experience, but also some empirical studies in computing education, suggests recurring misconceptions in the learning of Dijkstra's algorithm, such as misunderstandings about the meaning of distance labels, the role of permanently labeled nodes, and the logic behind iterative updates [11]. These observations indicate that learners may replicate procedural steps without consolidating the structural principles that justify them. Several teaching strategies have been proposed to mitigate these difficulties, following less traditional paths.

A first cluster of contributions relies on visualization and interactive digital tools, without involving physical manipulation: software-based environments present algorithm execution through interactive graphical interfaces. Animation-based systems and e-learning platforms for shortest path algorithms can support the step-by-step visualization of the algorithm, enabling students to observe dynamic updates of labels and real time edge relaxations in a controlled environment, with the graphical interfaces simulating algorithmic execution and providing automated feedback [1,7]. Game-oriented formats have also been explored in digital or symbolic settings. For example, structured question-driven designs can guide learners through decision sequences intended to mirror the logic of Dijkstra's algorithm [9], while modeling games in graph theory emphasize the translation of real world scenarios into network representations [2,6].

Although all these approaches encourage active cognitive engagement, the interaction remains mediated by screens or symbolic representations and does not involve tangible, physical manipulation of graph structures.

A second line of work introduces hybrid physical-digital environments, combining embodied interaction with computational support. For example, robot-assisted demonstrations with overlaid projections have been proposed to teach shortest path algorithms by coupling spatial movement with projected algorithmic information [5]. These systems leverage physical space to enhance engagement, yet remain dependent on technological infrastructure and do not necessarily emphasize the connection between algorithmic steps and optimization theory.

In contrast, the contribution by Jacob and Silvestri [4] appears more directly related to unplugged and tangible teaching approaches. The authors propose a mechanical device that simulates Dijkstra's algorithm through a physical apparatus, making the propagation of distance values observable without a computer. Their work shares with our proposal the goal of externalizing algorithmic dynamics into a manipulable artifact. However, some differences should be noted. First, the unplugged mechanical implementation is primarily designed for non-specialist or younger audiences, including primary school contexts, where the emphasis lies on intuitive exposure rather than on technical depth. Consequently, it neither aims at a formal analysis of key properties nor at connecting shortest path computation with linear programming duality. In contrast, our activity is embedded in university-level OR courses and explicitly integrates a primal-dual interpretation, using physical manipulation not only to illustrate procedural steps but also to anticipate formal dual reasoning. Second, our design relies exclusively on low-cost, easily retrieved materials, enabling individual use and scalability in large lecture settings without specialized mechanical devices.

Overall, existing literature confirms the pedagogical value of interactive and embodied approaches in teaching graph algorithms. The activity proposed here contributes to this landscape by combining tangible manipulation, algorithmic reasoning and an explicit bridge to duality, within a scalable framework suitable for large university classrooms.

3 Contents and Audience of OR Courses

The proposed activity is developed within the courses of Foundations of Operations Research, which are the first formal exposure to Optimization and OR. One course is offered in the third year of the Bachelor's degree in Automation and Control Engineering and the other in the first year of the Master's degree in Computer Engineering. Both last one semester (50 h, 5 ECTS) and aim to introduce the fundamental concepts of optimization, with emphasis on modeling and on the development of an algorithmic and structural "optimizer mindset."

The courses cover the main methodological pillars of Operations Research. After an introduction to modeling techniques (supported by a short laboratory component), graph theory is presented as a fundamental modeling framework for optimization problems. Undirected and directed graphs are introduced together with classical combinatorial problems such as minimum spanning tree, shortest paths, traveling salesman problem, maximum flow, and minimum-cost flow. For these problems, optimality conditions and the main solution algorithms are discussed, anticipating core ideas of duality and structural reasoning. The perspective is then generalized through Linear Programming, focusing on geometric interpretation, primal-dual relationships, optimality conditions, and complementary slackness. Integer Linear Programming is presented as a natural extension, with references to polyhedral methods and Branch-and-Bound. The course concludes with basic notions of computational complexity and an overview of general algorithmic paradigms (Greedy and Local Search), providing a conceptual framework for addressing computationally hard problems.

Despite large class sizes (approximately 220–300 students), the teaching approach is strongly oriented toward guided discovery and active learning. During lectures, students regularly work in small groups on short problem-based tasks. Multiple-choice questions and collective discussions are used to highlight key properties, critical aspects, and common misconceptions. Both correct and incorrect answers are analyzed, valuing error as a formative component of the learning process. This methodology supports conceptual understanding, especially in structurally rich domains such as graph theory.

The Bachelor's cohort is relatively homogeneous, while the Master's class is highly international and heterogeneous in background. In both cases, the scale and diversity of the audience make structured active-learning strategies particularly relevant for fostering engagement and deep understanding of graph-theoretical and combinatorial optimization concepts.

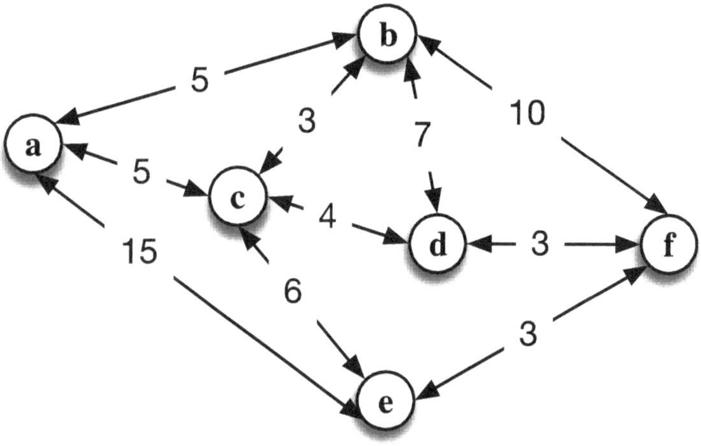

Fig. 1. Graph used in the physical example. Cord lengths are in accordance with the costs c_{ij}; a and f are the origin and destination nodes, respectively.

4 Material Preparation and Construction of the Physical Graph

The proposed activity, which has a duration of about 45 min, requires a bit of preparation at home for the students. The week before the lecture, the instructor provides a list of the material needed and instructions. In class, the material is used to construct a model of a graph: in the preceding lectures, the class has formalized, through an exercise, the mathematical model of the shortest path problem from an origin node s to a destination node t, which is reported in the next section for convenience. However, no information has been given about the properties of the solution, nor any algorithm has been illustrated.

Let us consider a simple directed graph with six nodes and ten bidirectional arcs. Figure 1 illustrates the example that we used.

The material needed for the activity includes six paper clips for the nodes, a few meters of non-stretchable rope, and a roll of masking tape to make labels with the value of c_{ij} on each edge. The students should cut the rope in small pieces to represent the bidirectional arcs. In order to have lengths that are not affected by the approximation errors, we suggest to use this rule to cut the rope: if the arc length is ℓ, the length of the rope is $3\ell + 2$ cm. The two additional centimeters are needed to clip the ropes together in the ending nodes. Thus, for the graph in Fig. 1 we need:

- six paper clips;
- 3 ropes with length 9+2 cm;
- 1 rope with length 12+2 cm;
- 2 ropes with length 15+2 cm;
- 1 rope with length 18+2 cm;
- 1 rope with length 21+2 cm;
- 1 rope with length 30+2 cm;
- 1 rope with length 45+2 cm.

Figure 2 shows the necessary material before carrying out the activity.

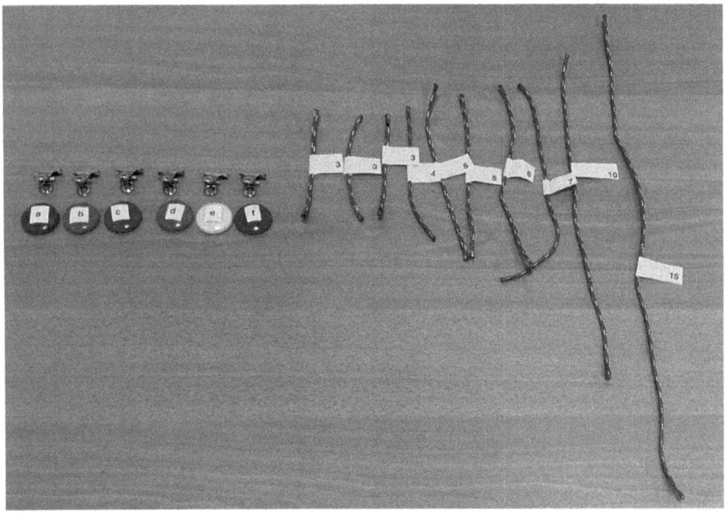

Fig. 2. Material preparation. Left: paper clips and colored magnetic lids with node labels; right: edges of the right length with cost labels.

5 Modeling the Problem and Dual Intuition

The activity aims at leading the students to formulate the dual of the shortest path problem without using the dualization rules, but rather as a natural modeling process from the emerging characteristics of the graph and their mutual relationships. Before starting the activity of assembling the graph provided in Fig. 3, we discuss the general formulation of the problem of finding the shortest path from node a to node f. A path is characterized by a sequence of arcs, therefore we introduce a set of binary variables x_{ij} for each arc (i, j) of the graph. Denoting by c_{ij} the length of arc (i, j), by A the set of arcs and by N the set of nodes, the model is:

$$\min \quad \sum_{(i,j)\in A} c_{ij}x_{ij} \tag{1}$$
$$\sum_{j:(a,j)\in A} x_{aj} = 1 \tag{2}$$
$$\sum_{i:(i,f)\in A} x_{if} = 1 \tag{3}$$
$$\sum_{j:(j,i)\in A} x_{ji} = \sum_{j:(i,j)\in A} x_{ij} \quad \forall i \in N \setminus \{a, f\} \tag{4}$$
$$x_{ij} \geq 0 \quad \forall (i, j) \in A. \tag{5}$$

Constraints (2) and (3) state that the path must start from node a and terminate in node f, while constraints (4) guarantee that if a path enters a node i different from the origin and the destination, it must also leave from it. Although this is a discrete optimization problem, enforcing integrality of the x_{ij} variables is unnecessary: the coefficient matrix is *totally unimodular*, which implies that any optimal solution is integer [12].

In the discussion of the mathematical model, we point out to the students that the model is not completely correct, as there exist feasible solutions where loops are included. However, under the assumption that arc costs are non-negative, the risk of loops is neutralized by the objective function which must be minimized. The activity proceeds to the actual construction of the physical model of the graph using the material that has been prepared in advance. The result is depicted in Fig. 3.

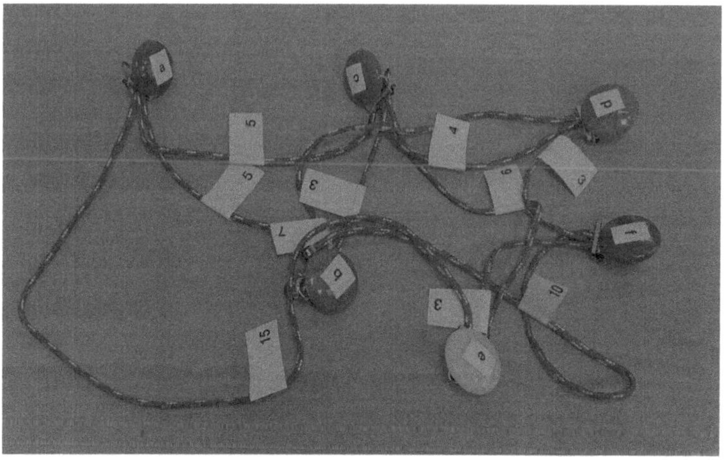

Fig. 3. Construction of the physical model.

It is important that each student/group of students has a physical model of the graph so that they can manipulate it themselves rather than watching the instructor. Once the physical models of the graph have been assembled, the instructor invites the students to make use of the models to compute the shortest path from node a to node f. After the first moments when the students are quite cautious in touching their models, someone gets the intuition, even though not yet substantiated by knowledge of the problem's dual, of taking node a with one hand, node f with the other hand and pulling them apart (see Fig. 4).

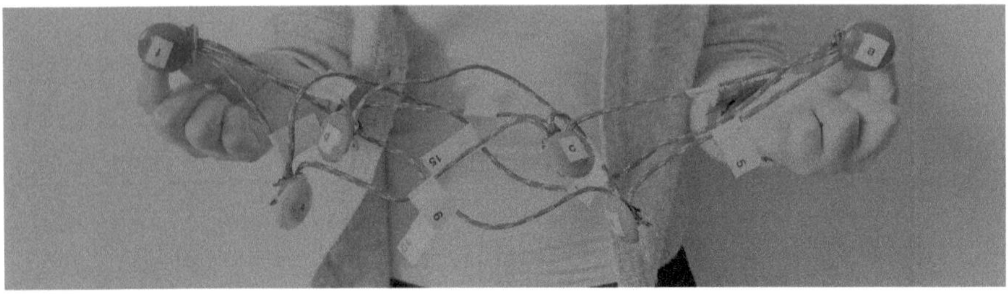

Fig. 4. Solution method: separate the extremes as much as possible without disconnecting any rope.

The class soon convenes that the stretched ropes define the sequence of arcs belonging to the shortest path, an observation made easy and verifiable by the simplicity of the example. However, when the instructor asks what type of optimization has been done by pulling apart the extremes of the path, some perplexities emerge. The doubts are more evident when the instructor asks whether the operation that brought to the shortest path is a minimization or a maximization.

The discussion that follows has the final intent of defining the actual problem that has been solved using the model: the operation is maximizing the distance between the extremes of the path, while the decisions are not about which arcs to select, but they are about the relative distance between adjacent nodes, and the physical restrictions that must be enforced aim at avoiding to destroy the model, i.e., avoiding separation of the nodes by pulling them at a distance greater than the length of the arc joining them.

Based on the outcome of this discussion on the physical restrictions on each node pair, the class is ready to formulate the problem. The decision variables π_i are associated to each node $i \in N$ and represent the *level*, or potential, of the node. The objective function is:

$$\max \pi_f - \pi_a. \tag{6}$$

The physical restrictions lead to constraints, one per arc, stating that the difference between the level of the head node j and that of the tail node i cannot exceed the length of the rope:

$$\pi_j - \pi_i \leq c_{ij} \qquad \forall (i,j) \in A. \tag{7}$$

When in the course we deal with Linear Programming and Duality theory, a few weeks later, we point out that, after changing the sign to one equation, the shortest path problem and the "rope problem" define a pair of dual problems. However, for the time being, we simply observe that the arcs selected in the shortest path are those such that the difference of their levels is equal to the arc length: in other words, the "tight" arcs, which would be broken if pulled further, are those belonging to the shortest path, while the "loose" arcs correspond to inactive constraints. This turns out to be the complementary slackness property, linking the primal and the dual solutions: the property will properly be recalled and explained later, when dealing with Linear Programming. Moreover, we point out that both the mathematical and the physical models have one degree of freedom, since the solution is independent from the value of one of the levels. Indeed what we optimize is the difference of the levels between the extremes of the path, without fixing the level of the starting node to a given value.

6 Using the Physical Graph as a Guide to Derive Dijkstra's Algorithm

In the activity described above, we exploited a physical model to derive the optimal solution of the problem. In obtaining the optimal solution, all decision variables of the model (i.e., node heights) are optimized *simultaneously*.

However, if we want to describe the algorithm resulting from the *sequential* version of the method that we used to determine the solution, we must take into account that each iteration considers a different state of the graph. In our activity we do so by starting with a *decomposed* physical graph, where nodes are detached from one another and each node only has ropes from its *forward star*, i.e., the set of arcs emanating from the node.

When solving the problem through the model, we used two hands to pull apart the extremes of the path. Now the intuition to follow the steps of a sequential algorithm is to use one hand only, moving one node at a time. Instead of using the horizontal dimension, we use the vertical one. We will explain the steps of the manual procedure mapping them according to the algorithm's description provided in Algorithm 1. With the help of a volunteer, or by sticking a magnet to the clips and using a magnetic board, as a first step we raise the origin node a in a high position (Step 2 in Algorithm 1), which we indicate as "level 0". All other nodes remain at the base of the board, which one could denote as "level infinity". Only the three ropes incident with a are still in place, though the other extremes are missing (see Fig. 5a).

The objective of the procedure is to join the detached components by raising one node at a time. The node to be raised, due to gravity, will be at such level that the connecting arc is stretched. Moreover, we select the node among those in L, i.e., the set of nodes that can be raised, which guarantees being raised maximally, or equivalently being placed at the "lower level" (Step 5), so that its position will not be changed in the subsequent steps. Consequently, we update the state of its forward star (the for loop at Step 7).

In the second iteration, two dangling ropes from node a have the same length, so we can select either node b or node c—in this case, b is selected, see Fig. 5b. We proceed in the same way by raising node c (Fig. 5c), then d and e (Figs. 5d and 5e respectively) until we set the height of the last node f as in Fig. 5f.

Algorithm 1: Dijkstra's algorithm

```
   input: Graph (N, A); travel cost c_ij; origin and destination nodes s, t
   output: Predecessors p : N → N, potentials π : N → ℝ
1  p_s ← s; p_i ← ∅ for all i ∈ N \ {s}
2  π_s ← 0; π_i ← +∞ for all i ∈ N \ {s}
3  L ← {s}
4  while L ≠ ∅ do
5      Select i ∈ L such that π_i = min_{j∈L} π_j          // Set level of node i
6      L ← L \ {i}
7      for (i, j) ∈ δ+(i) do
8          if π_j > π_i + c_ij then
9              π_j ← π_i + c_ij    // Update label of end nodes of dangling rope
10             p_j ← i
11             L ← L ∪ {j}
12 Return p, π
```

We can note the direct correspondence between height of the nodes, variables π_i, and the labels of Dijkstra's algorithm. This implies that, at each iteration, we will select the node whose label will be fixed at its minimum value: we enforce, through a physical model, the property that the rope connecting the node with the above graph is stretched. Eventually, the correctness of the algorithm directly follows from the property of the physical model. Even though not all ropes have been joined, the ones that are not dangling are stretched and identify the arcs belonging not only to the shortest path from the origin to the destination, but also to the shortest path tree having root in the origin.

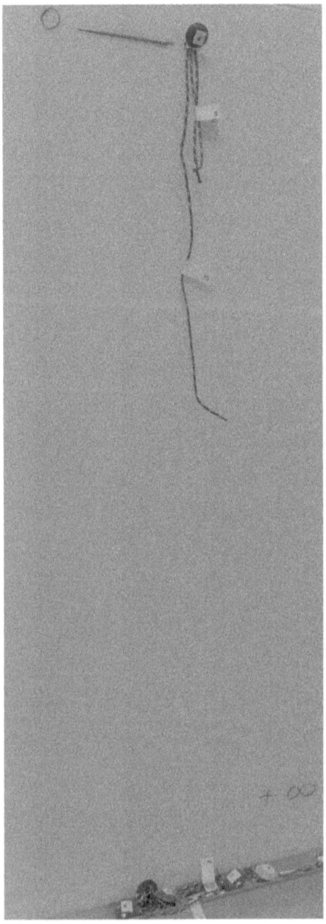

(a) Step #1: node *a* placed at level 0

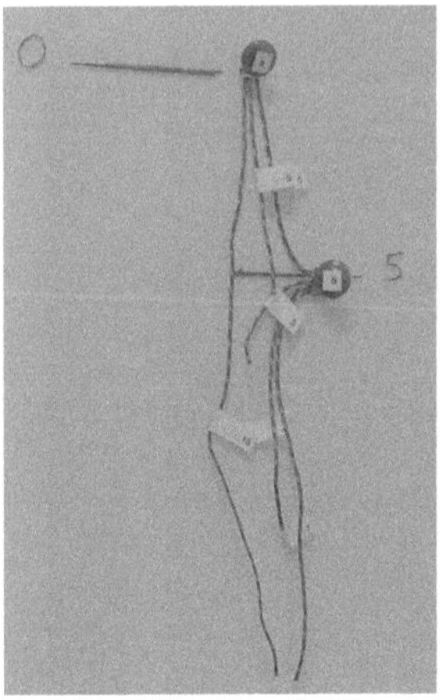

(b) Step #2: node *b* is attached with cord *ab* at level 5

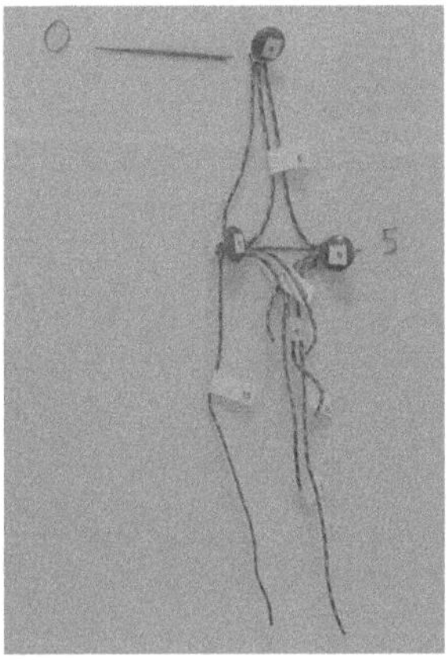

(c) Step #3: node *c* attached at level 5

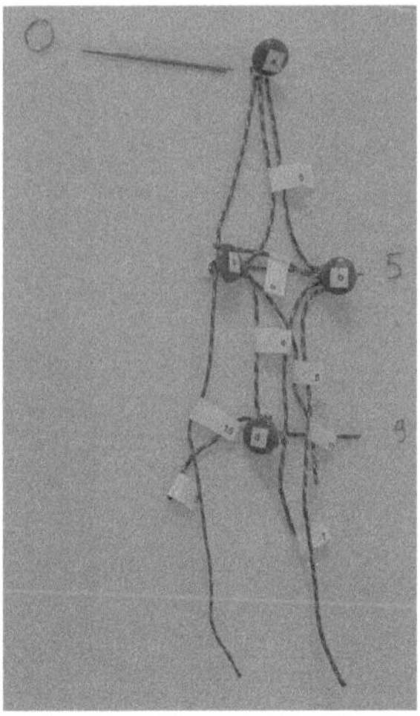

(d) Step #4: node *d* is attached

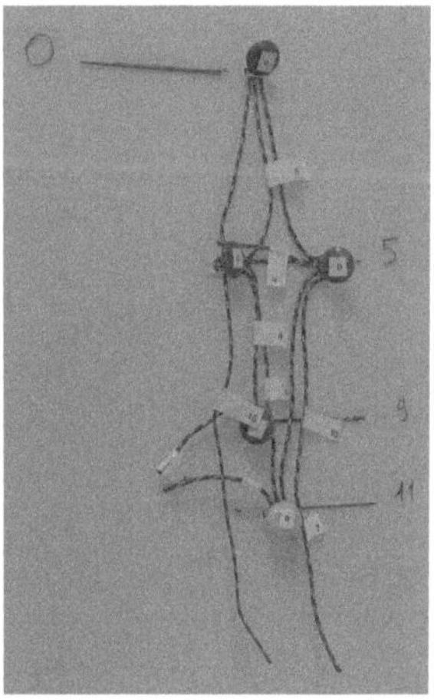

(e) Step #5: node *e* is attached

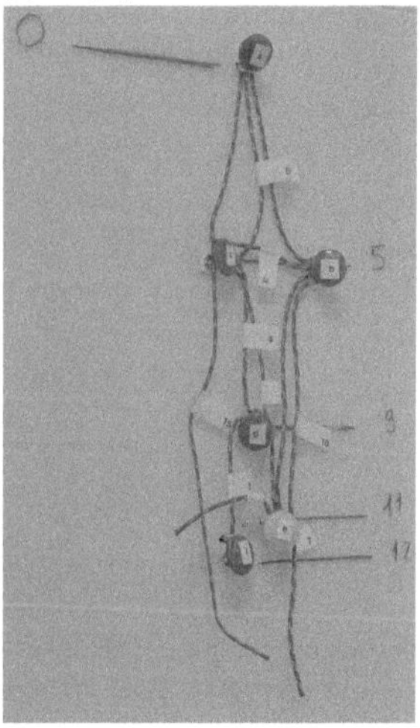

(f) Final step: node *f* is reached

Fig. 5. Step-by-step execution of Dijkstra's algorithm.

7 Concluding Remarks

This paper has presented a structured classroom activity for introducing the shortest path problem and Dijkstra's algorithm within introductory Operations Research courses. Through the construction and guided manipulation of a physical weighted graph, the activity allows students to experience shortest path computation before formalization and to progressively uncover its dual interpretation in terms of node potentials and feasibility constraints. The same tangible model is then used to mirror the sequential logic of Dijkstra's algorithm, making explicit the connection between physical actions, tentative labels, and optimality conditions. In this way, the activity aims to bridge procedural execution and structural reasoning, anticipating formal concepts through direct experience.

The main strengths of the approach lie in its conceptual coherence, easy implementation and scalability. The materials are inexpensive and easily retrieved, making the activity suitable even for large cohorts, while still ensuring that each student can actively participate. Moreover, the design explicitly connects graph-theoretic intuition, algorithmic logic and optimization theory within a single learning sequence, supporting a smoother transition to later formalization.

Over more than ten years of implementation in undergraduate and graduate courses, we have gathered consistent qualitative evidence of its effectiveness. Although we have not conducted controlled quantitative assessments, classroom experience suggests significant levels of participation during lectures, stronger engagement across diverse student profiles and a higher level of comprehension and retention. Students frequently report that the activity clarifies the logic of the algorithm and helps them better understand the relationship between shortest paths and dual reasoning. From an instructional perspective, the additional effort required to design and manage such activities has proven worthwhile in light of the involvement and improved conceptual clarity observed in practice.

References

1. Borissova, D., Mustakerov, I.: E-learning tool for visualization of shortest paths algorithms. Trends J. Sci. Res. **2**, 84–89 (2015)
2. Cordero, A., Jordán, C., Murillo-Arcila, M., Sanabria-Codesal, E.: A game for learning how to model in graph theory. Mathematics **10**(12), 1969 (2022)
3. Freeman, S., et al.: Active learning increases student performance in science, engineering, and mathematics. Proc. Natl. Acad. Sci. **111**(23), 8410–8415 (2014). https://doi.org/10.1073/pnas.1319030111
4. Jacob, R., Silvestri, F.: Unplugging Dijkstra's algorithm as a mechanical device. In: Creative Mathematical Sciences Communication: 7th International Conference, CMSC 2024, Trier, Germany, 7–10 October 2024, Proceedings, pp. 39–49. Springer, Heidelberg (2024). https://doi.org/10.1007/978-3-031-73257-7_3
5. Jolakoski, P., Deja, J.A., Čopič Pucihar, K., Kljun, M.: Teaching shortest path algorithms with a robot and overlaid projections (2024). https://arxiv.org/abs/2411.15535

6. Jordán, C., Murillo-Arcila, M., Torregrosa, J.R.: The STEM methodology and graph theory: some practical examples. Mathematics **9**(23) (2021)
7. Liang, Y.D.: Teaching and learning graph algorithms using animation. J. Comput. Sci. Educ. **9**(2), 23–29 (2018)
8. Maffia, A., Pelillo, M.: Intuitive geometry by Emma Castelnuovo: still contemporary in the digital devices' era. Exp. Teach. Math. Sci. Technol. - EDiMaST **1**(2), 131–140 (2015)
9. Ramle, R., Rosli, D., Nathan, S., Berahim, M.: Question-led approach in designing Dijkstra algorithm game-based learning: a pilot study. Int. J. Eval. Res. Educ. (IJERE) **9**, 926 (2020)
10. Theobald, E.J., et al.: Active learningnarrows achievement gaps for underrepresented students in undergraduate science,technology, engineering, and math. Proc. Natl. Acad. Sci. **117**(12), 6476–6483 (2020). https://doi.org/10.1073/pnas.1916903117
11. Tilanterä, A., Sorva, J., Seppälä, O., Korhonen, A.: Students struggle with concepts in Dijkstra's algorithm. In: Proceedings of the 2024 ACM Conference on International Computing Education Research, ICER 2024, vol. 1, pp. 154–165. Association for Computing Machinery, New York (2024)
12. Nemhauser, G.L., Wolsey, L.A.: Integer and Combinatorial Optimization. Wiley (2014). https://doi.org/10.1002/9781118627372

Data Stream Algorithms Unplugged

Philip Bille(✉) and Inge Li Gørtz

Technical University of Denmark, Kgs., Lyngby, Denmark
{phbi,inge}@dtu.dk

Abstract. Data stream algorithms are an advanced topic in computer science that focus on processing massive data sets arriving as an online stream of items. The key algorithmic challenge is to process the items online in one or a few passes while only using very limited working space. Data stream algorithms are typically simple and elegant, but the mathematics behind their performance guarantees or correctness proofs often involve intricate combinatorial or probabilistic arguments. For many students, this is a significant barrier to effective learning and intuition-building in the area. In this paper, we address this challenge and develop an unplugged activity that introduces data stream algorithms. The activity consists of a simple, unplugged, physical simulation of data stream algorithms using Lego bricks drawn one by one from an opaque bag and placed on a table or discarded according to the rules of the specific algorithm. In this model, we present a simulation of the classical Misra-Gries algorithm [*Sci. Comput. Program.*, 1982]. As a key feature, we provide an explicit, elementary proof of correctness that relies on the configuration of the Lego bricks assembled during the simulation. The proof is sufficiently simple that most students (with or without a few hints) can discover it on their own during a short group exercise. The activity can be easily adapted to different audiences, depending on their computer science and mathematical backgrounds (including high school students and computer science students), and we discuss these variants.

Keywords: Data Stream Algorithms · CS unplugged · Frequent Items

1 Introduction

Computer Science Unplugged (CS Unplugged) is a collection of activities that teach computer science without using computers or other digital technologies. Instead, CS Unplugged uses engaging demonstrations, puzzles, and games, as well as physical props such as cards, strings, paper, pens, and pencils, to explain concepts. CS Unplugged was popularized by Bell et al. in their book [7], which presents a collection of activities on fundamental algorithms for primary school children. Since then, numerous activities across a wide range of computer science topics and age groups have been designed, see e.g., [2,8,21,24,26,32–36], surveys [5,6], and the homepage www.csunplugged.org.

Data stream algorithms are an advanced topic in computer science that focuses on processing massive streams of data arriving online. Typical examples

M. Löffler et al. (Eds.): CMSC 2026, LNCS 16951, pp. 16–26, 2027.
https://doi.org/10.1007/978-3-032-38887-2_2

include analyzing network traffic, online auctions, bank transactions, communication logs, atmospheric data, and astronomical data. The key features are that algorithms must process the input data stream in one or a few passes and that their working space is limited, ideally much smaller than the input data stream. This captures the online, massive data-processing scenario where we cannot store the input, and the only feasible access to the data is by making a few passes over it. Data stream algorithms date back to the late 1970s and the topic is now an established and well-studied area with numerous results for a wide range of problems, see e.g., [1,4,9–15,19,22,23,27–30] and surveys [3,17,20,31].

In academic education, data stream algorithms are typically taught in advanced algorithms courses and require significant prerequisites in both computer science and mathematics. While most data stream algorithms are simple, elegant, and easy to learn, the mathematics behind their performance guarantees or correctness proofs often involve intricate combinatorial or probabilistic arguments. For many students, this is difficult and poses a significant barrier to effective learning and intuition-building in this area. Thus, a natural question is whether we can develop educational material to remedy this.

1.1 Contributions

Our goal in this paper is to address the above challenge and develop a new approach to introduce students to data stream algorithms and to better support their intuition in this area. To do so, we describe an unplugged activity that includes a physical simulation of the data stream model, a central algorithm in the model, and a simple, intuitive, and elementary proof of correctness that relies on the simulation's features. Specifically, we make the following contributions:

- We present a simple, unplugged model of data stream algorithms that uses Lego bricks of different colors, drawn one by one from an opaque bag and placed on a table or discarded according to the algorithmic rules. The "stream" of Lego bricks from the bag represents the online processing of items, and the table represents the algorithm's working memory.
- We consider the *k-frequent items problem*, that is, given an integer parameter k, we want to compute the items that appear with frequency more than n/k in a data stream of n items. In our unplugged model, this corresponds to computing which colors appear with frequency more than n/k.
- We implement the classical Misra-Gries algorithm [28] for the k-frequent items problem in the unplugged model. We show how to represent the algorithm's internal state using simple stacks of bricks. In addition, we maintain a special discard pile of groups of bricks that are clicked together and are not part of the algorithm's internal state but only used for analysis. We show how to use the discard pile to give a simple, intuitive, and elementary proof of correctness.
- Our activity can be easily adapted to different target audiences (including high school and computer science students), depending on their mathematical and computer science backgrounds and interests. We discuss several extensions and variants of our activity to accommodate these groups.

1.2 Outline

We first discuss the algorithmic background for the data stream model and the Misra-Gries algorithm in Sect. 2. We present the unplugged model and algorithms in Sect. 3. We then discuss how to adapt the activity to different target audiences in Sect. 4. Finally, we conclude and discuss open problems in Sect. 5.

2 Algorithmic Background

We discuss the algorithmic background of the data stream model, the frequent items problem, and its classical solutions.

2.1 Data Stream Model and the Frequent Items Problem

In the data stream model, the input is accessed sequentially as a stream $\mathcal{X} = x_1, \ldots, x_n$ of n items, and the goal is to process them using a working memory that is small compared to the length n of the stream and make as few passes over the input as possible. Ideally, we want to perform a single pass over the input and use space polylogarithmic in the length of the stream.

The *frequency* of an item x in $\mathcal{X}$, denoted f_x, is the number of copies of x in $\mathcal{X}$. An item x is *k-frequent* if $f_x > n/k$. Given a stream $\mathcal{X}$ and an integer parameter $k \geq 1$, the *k-frequent items problem* is to compute the k-frequent items in $\mathcal{X}$. If $k = 2$, we call the problem the *majority problem.*

2.2 Misra-Gries Algorithm

Misra-Gries [28] proposed a simple space-efficient two-pass solution that work as follows:

Pass 1: We maintain a table C of at most $k-1$ entries. Each table entry stores an item x with a count, denoted $C[x]$. Initially, C is empty. For each item x_i in $\mathcal{X}$,

- If $x_i \in C$, we increment the corresponding counter $C[x_i]$.
- If $x_i \notin C$ and C contains less than $k-1$ items, we insert x_i into C and set $C[x_i] = 1$.
- If $x_i \notin C$ and C contains $k-1$ items, then for each $x \in C$, we decrement $C[x]$ and delete the entry if $C[x]$ becomes 0.

Pass 2: Let C^* be the contents of the table C at the end of pass 1. We compute the frequency of each item in C^* by maintaining a counter for each item in C^* and incrementing the corresponding counter each time we see an element from C^* in $\mathcal{X}$. Finally, we output the items from C^* that are k-frequent.

The algorithm uses $O(k)$ space to store the table C, and the time is dominated by a constant number of dictionary operations for each item. With additional techniques [16,25], these can be implemented in worst-case constant time. To show correctness, the main observation is that all k-frequent items in must be in C^* at the end of the first pass.

Lemma 1. *If an item x is k-frequent in $\mathcal{X}$, then $x \in C^*$ after the first pass.*

Proof. Consider the items in C^* and let $c = \sum_{x \in C^*} C[x]$ be the total count of them. We can partition all items in $\mathcal{X}$ into two sets: c items in C^* and $(n-c)/k$ groups such that each group contains k distinct items (each such group corresponds to a decrement operation in the third case of the algorithm). For contradiction, suppose there is another k-frequent item $x' \notin C^*$. Since the second set contains groups of k distinct items, it can have at most $(n-c)/k$ copies of x'. Thus, the first set must also contain at least one copy of x', contradicting the fact that $x' \notin C^*$.

The above proof is adapted from the original paper by Boyer and Moore [10] and extended to general k. Intuitively, the idea is that k distinct items can always be discarded without affecting the set of k-frequent items. A wide range of similar "grouping" arguments appear in other proofs [16,18,25,28]. Unfortunately, we observe that for many students, such arguments are abstract, difficult to grasp, and hard to build a solid intuition for.

3 Unplugged Data Stream Algorithms and Frequent Items

We now present our new unplugged data-stream algorithm setup with Lego bricks, the k-frequent items problem in this setup, and our implementation of the Misra-Gries algorithm. We first present a generic version of our activity, then discuss how to adapt it for different target audiences in the following section.

3.1 Data Stream Algorithms with Lego Bricks

We need the following items for our unplugged, physical demonstration of data stream algorithms:

- A set of around 15–20 Lego bricks of different colors. The bricks will represent the items in the stream, and bricks of the same color represent copies of the same item. The set of bricks should include subsets with varying color frequencies to illustrate the algorithm's different outcomes. We recommend using the larger Duplo bricks, which are easily visible in larger classrooms, but any bricks that click together will suffice.
- An opaque bag that will hold the bricks. To simulate a pass of a data stream, we place our (chosen subset of) bricks into the bag and pull them out of the bag one by one. Thus, students will only see a single brick at a time and cannot see any future bricks.

- A table with a stable surface that is clearly visible to all students. We use this to illustrate the algorithm's behavior and analyze it by clicking together Lego bricks as we pull them out of the bag. Effectively, the table will serve as the algorithm's working memory.

This setup supports the main features of data stream algorithms, i.e., we process items online (pulling them out of the bag one by one) and have limited working memory (the table). To implement multiple passes, we may return the bricks to the bag to "restart" the stream. We note that the setup is simple and sufficiently general for our purposes, but it is not suited for all types of streaming algorithms. For instance, we can only deal with a limited number of distinct items (the available brick colors), and items are permuted randomly between passes (shuffled in the bag).

3.2 Frequent Items with Lego Bricks

We now show how to solve the k-frequent items problem in our Lego brick setup. We recommend demonstrating the algorithms with small k such as 2 (the majority problem), 3, or 4. We will represent the actions of our algorithms by building stacks of Lego bricks on the table using bricks pulled from the bag. Each stack consists of bricks of the same color, and the algorithm's working space will be represented by the total number of stacks. Thus, the k-frequent items problem in our setting is to compute the colors that appear more than n/k times in the stream of bricks, while using as few stacks on the table as possible and as few passes as possible.

3.3 Simple Solution

As a warm-up and to build intuition for the problem, we first present the naive single-pass solution to the audience. As discussed, we will maintain stacks of bricks of the same color. Initially, all stacks are empty (the table is clear). For each brick x_i we pull out of the bag,

- If x_i matches a stack, we click x_i to the top of that stack.
- If x_i does not match any stack, we make a new stack consisting of x_i.

Note that the stack heights represent the frequency of each color. At the end, we inspect the stacks and check whether any stack contains more than n/k bricks, reporting them as k-frequent items.

Clearly, the naive solution correctly solves the problem. The main issue is that we need a stack on the table for each distinct color that appears in the stream. We explain to the students that the key challenge is to avoid these stacks, and our goal is to solve the problem with significantly fewer stacks.

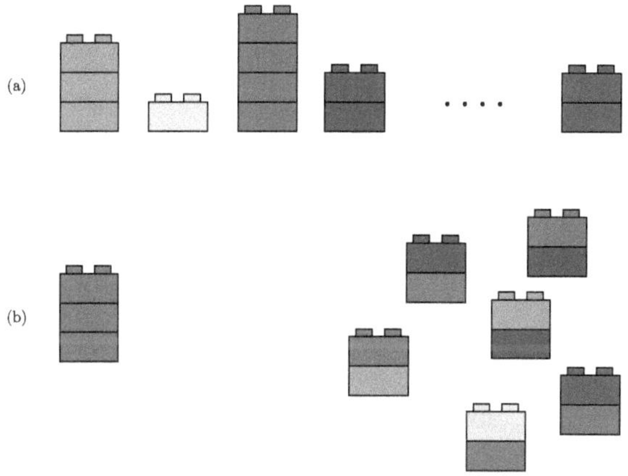

Fig. 1. (a) Stacks during the simple solution. (b) Stack and discard pile of the Misra-Gries algorithm with $k = 2$ after pass 1. There are 8 red bricks out of the 15 total, so red is the majority. We can have at most 1 red brick in each pair in the discard pile that can contain at most 7 pairs. Thus, at least 1 red brick must be in the stack. (Color figure online)

3.4 Misra-Gries Algorithm

We then present the Misra-Gries algorithm.

Pass 1: We maintain at most $k-1$ stacks of bricks on the table. Each stack will contain bricks of the same color. Initially, all stacks are empty (i.e., the table is clear). For each brick x_i, we pull out of the bag,

- If x_i matches a color of a stack, we click x_i to the top of that stack.
- If x_i does not match any stack color and there are fewer than $k-1$ stacks, we create a new stack consisting of x_i.
- If x_i does not match any stack color and there are $k-1$ stacks, we click off the top of each stack. We then click the resulting top bricks and x_i together to form a group of k bricks, and put this group in a *discard pile*. For dramatic effect, we can throw these in a trash can or on the floor.

Pass 2: Unclick all the bricks, put them back in the bag, and remember only the colors C^* that were in the stacks at the end of pass 1. Start a new pass, and for each brick x_i we pull from the bag,

- If x_i matches a color in C^*, we click x_i to the top of that stack.
- If x_i does not match any color in C^* we ignore it.

At the end, we inspect the stacks, check whether each stack has a height greater than n/k, and report the colors of those that do as the k-frequent items.

We discuss the algorithm with the audience and, if needed, repeat the demonstration using different subsets of bricks to clarify its behavior and illustrate different outcomes. We then further discuss to what extent the algorithm satisfies the stated goals. We observe that it uses two passes and no more than $k - 1$ stacks at any point in time. Thus, it remains to argue that it correctly solves the problem, and this will be the main focus of our discussion with the audience. The key property that we want to show is Lemma 1 adapted to our setting:

Lemma 2. *Any color that appears more than n/k times must be the color of a stack at the end of pass 1.*

We argue that this property must imply that the algorithm correctly solves the problem. Then we invite the audience to discuss why this property is true, either individually or in small groups. We recommend giving the audience about 5 min for this task. While doing so, we leave the stacks and the discard pile after a demonstration of pass 1 visible for the students. If needed, we provide the following hint: if a color appears more than n/k times, where can the bricks of that color be at the end of pass 1? And further: look closely at the discard pile. In our experience, most of our target audiences will discover a correct argument for Lemma 2 by realizing that since each group in the discard pile all have distinct colors, a frequent color must appear in a stack after pass 1 (see Fig. 1 for an illustration). More formally:

Proof. At the end of pass 1, bricks of a given color must be either in a stack or in the discard pile. The discard pile consists of at most n/k groups of k bricks. Each group of k bricks consists of bricks of k distinct colors. Hence, there can be at most n/k bricks of any given color in the discard pile. Therefore, if a color appears *more* than n/k times, that color must be in a stack at the end of pass 1.

Note that the discard pile helps to make the "grouping" argument more explicit since we can now physically point to the grouping.

4 Target Audiences

The above generic activity can be adapted to different audiences. We discuss our experiences with some of these below.

4.1 Late Primary School/High School Students

For this audience, our activity is best suited for demonstrating computational thinking and introducing students to clever algorithmic ideas that require non-trivial combinatorial insights. Thus, we focus on the topic mainly as an algorithmic puzzle and ignore programming details. To motivate the problem, we explain and discuss the importance of data streaming algorithms in real-world scenarios.

For this audience, we focus on the majority problem ($k = 2$). If possible, provide each group with a set of Lego bricks to help them test their ideas.

Before we let the students discuss why Lemma 2 is true, we suggest to first let the students "play" with the proof of the algorithm by trying to give the bricks in an order that would break it, and then use the insight obtained to understand why the proof is correct. As a further exercise, we suggest that students generalize to larger k (starting with $k = 3$) using their own set of bricks.

4.2 Graduate Algorithms

For this audience, the main use is to introduce students with a background in algorithms to the data stream model and its main challenges. We motivate the problem as in Sect. 4.1. We focus on general k and the proof of Lemma 2. Afterward, we relate the unplugged model and Misra-Gries algorithm to the standard (plugged) data stream model and Misra-Gries algorithm.

Furthermore, we propose a set of advanced exercises with elegant solutions using the unplugged model.

- Let f_i be the frequency of color i in the stream $\mathcal{X}$. Prove that the estimated frequency $\hat{f}_i$ of color i is at least $f_i - \frac{n-c^*}{k}$, where c^* is the sum of the values of the counters in C^*. *Solution: the total number of bricks in the discard pile is at most $n - c^*$. Thus, each color can occur at most $\frac{n-c^*}{k}$ times in the discard pile.*
- Give an example of a stream that illustrates why the algorithm cannot be used to compute colors with frequency *at least* n/k (instead of *more than* n/k). *Solutions: There are several solutions to this. E.g., a stream of bricks of k colors that each occur n/k times. All bricks of the kth color are at the end of the stream. This will result in all stacks being empty at the end of the stream.*
- Assume you have two streams $\mathcal{X}_1$ and $\mathcal{X}_2$ and run the Misra-Gries algorithm on each of them independently with the same k. Now merge the two results as follows. If a color has a stack in both streams, add the stacks together. You now have at most $2(k-1)$ stacks. Let h be the value of the kth largest of these stacks. Remove a brick h from each stack, keeping only the non-empty stacks. Prove that any color that appears more than n/k times must be a color of a non-empty stack. *Solution: Merge the two discard piles into one. All the groups in the merged discard pile have the same property as before. When removing h bricks from each stack, do it in h rounds. In each round, remove a brick from each non-empty stack, click them together, and put them in a new discard pile. The important observation here is that each newly created group has at least k bricks, and the bricks in a group have distinct colors. Thus, a color cannot be discarded more than n/k times.*

As above, we suggest providing each group with a set of Lego bricks to help them test their ideas.

4.3 Public Outreach

The unplugged activity is also suitable for more popular talks aimed at promoting computer science ideas and computational thinking. In this scenario, we use the same version of the activity as in Sect. 4.1. However, we emphasize the connection to real-world practical problems.

5 Conclusion and Open Problems

We presented an unplugged activity for introducing data stream algorithms through a physical simulation with Lego bricks. The activity captures the central constraints of the streaming model—online processing and limited memory—while making the Misra-Gries algorithm and its correctness proof concrete and visually accessible. In particular, the discard pile provides a tangible representation of the grouping argument, allowing students to discover the key invariant themselves rather than encountering it only as an abstract combinatorial proof. The activity is simple to run, requires minimal materials, and can be adapted to audiences ranging from high school students to graduate algorithms students and public outreach settings.

Several natural open problems remain. First, it would be valuable to design and evaluate unplugged activities for other core data streaming techniques, such as Count-Min, sampling-based methods, or frequency-moment estimators. Second, a systematic classroom study could measure how the physical grouping argument affects student understanding compared with a traditional lecture-based presentation. Finally, it would be interesting to explore whether the unplugged model can support deeper advanced topics while preserving the simplicity and intuition that make the activity effective.

Acknowledgments. We thank Gad M. Landau for the initial inspiration for this work. This work was partially funded by the Danish Research Council (grant numbers 10.46540/3105-00302B and 10.46540/4283-00129B)

Disclosure of Interests. The authors declare that they have no conflict of interest.

References

1. Alon, N., Matias, Y., Szegedy, M.: The space complexity of approximating the frequency moments. In: Proceedings of the 28th STOC, pp. 20–29 (1996)
2. Angara, P.P., Stege, U., MacLean, A., Müller, H.A., Markham, T.: Teaching quantum computing to high-school-aged youth: a hands-on approach. IEEE Trans. Quantum Eng. **3**, 1–15 (2021)
3. Babcock, B., Babu, S., Datar, M., Motwani, R., Widom, J.: Models and issues in data stream systems. In: Proceedings of the 21st PODS, pp. 1–16 (2002)
4. Bar-Yossef, Z., Jayram, T.S., Kumar, R., Sivakumar, D.: An information statistics approach to data stream and communication complexity. J. Comput. Syst. Sci. **68**(4), 702–732 (2004)
5. Battal, A., Adanır, G.A., Gülbahar, Y.: Computer science unplugged: a systematic literature review. J. Educ. Tech. Sys **50**(1), 24–47 (2021)

6. Bell, T., Rosamond, F., Casey, N.: Computer science unplugged and related projects in math and computer science popularization. In: The Multivariate Algorithmic Revolution and Beyond: Essays Dedicated to Michael R. Fellows on the Occasion of his 60th Birthday, pp. 398–456 (2012)
7. Bell, T., Witten, I.H., Fellows, M.R.: Computer Science Unplugged - An Enrichment and Extension Programme for Primary-Aged Children (2002)
8. Benavides-Escola, C.A., Martín-Barroso, E., Zapata-Cáceres, M., Román-González, M.: Improvement of computational thinking skills through unplugged activities in upper secondary education. In: Proceedings of the 19th WiPSCE (2024)
9. Bhuvanagiri, L., Ganguly, S., Kesh, D., Saha, C.: Simpler algorithm for estimating frequency moments of data streams. In: Proceedings of the 17th SODA, pp. 708–713 (2006)
10. Boyer, R.S., Moore, J.S.: MJRTY—a fast majority vote algorithm. In: Automated Reasoning: Essays in Honor of Woody Bledsoe, pp. 105–117 (1991)
11. Charikar, M., Chen, K., Farach-Colton, M.: Finding frequent items in data streams. In: Proceedings of the 29th ICALP, pp. 693–703 (2002)
12. Charikar, M., O'Callaghan, L., Panigrahy, R.: Better streaming algorithms for clustering problems. In: Proceedings of the 35th STOC, pp. 30–39 (2003)
13. Coppersmith, D., Kumar, R.: An improved data stream algorithm for frequency moments. In: Proceedings of the 15th SODA, pp. 151–156 (2004)
14. Cormode, G., Muthukrishnan, S.: An improved data stream summary: the count-min sketch and its applications. J. Algorithms **55**(1), 58–75 (2005)
15. Cormode, G., Muthukrishnan, S.: What's hot and what's not: tracking most frequent items dynamically. ACM Trans. Database Syst. **30**(1), 249–278 (2005)
16. Demaine, E.D., López-Ortiz, A., Munro, J.I.: Frequency estimation of internet packet streams with limited space. In: Proceedings of the 10th ESA, pp. 348–360 (2002)
17. Demetrescu, C., Finocchi, I.: Algorithms for data streams. In: Handbook of Applied Algorithms: Solving Scientific, Engineering and Practical Problems, pp. 241–269. Wiley Online Library (2008)
18. Fischer, M.J., Salzberg, S.L.: Finding a majority among n votes: solution to problem 81-5. J. Algorithms **3**(4), 376–379 (1982)
19. Frahling, G., Indyk, P., Sohler, C.: Sampling in dynamic data streams and applications. In: Proceedings of the 21. SoCG, pp. 142–149 (2005)
20. Gibbons, P.B., Matias, Y.: Synopsis data structures for massive data sets. In: External Memory Algorithms, DIMACS Series in Discrete Mathematics and Theoretical Computer Science, vol. 50, pp. 39–70 (1999)
21. Greenberg, R.I.: Educational magic tricks based on error-detection schemes. In: Proceedings of the 22nd ITiCSE, pp. 170–175 (2017)
22. Indyk, P.: Algorithms for dynamic geometric problems over data streams. In: Proceedings of the 36th STOC, pp. 373–380 (2004)
23. Indyk, P., Woodruff, D.: Optimal approximations of the frequency moments of data streams. In: Proceedings of the 27th STOC, pp. 202–208 (2005)
24. Jacob, R., Silvestri, F.: Unplugging Dijkstra's algorithm as a mechanical device. In: Proceedings of the 7th CMSC, pp. 39–49 (2024)
25. Karp, R.M., Shenker, S., Papadimitriou, C.H.: A simple algorithm for finding frequent elements in streams and bags. ACM Trans. Database Syst. **28**(1), 51–55 (2003)

26. Lehner, L., Landman, M.: Unplugged decision tree learning–a learning activity for machine learning education in K-12. In: Proceedings of the 7th CMSC, pp. 50–65 (2024)
27. Metwally, A., Agrawal, D., El Abbadi, A.: Efficient computation of frequent and top-k elements in data streams. In: Proceedings of the 10th ICDT, pp. 398–412 (2005)
28. Misra, J., Gries, D.: Finding repeated elements. Sci. Comput. Program. **2**(2), 143–152 (1982)
29. Morris, R.: Counting large numbers of events in small registers. Commun. ACM **21**(10), 840–842 (1978)
30. Munro, J.I., Paterson, M.S.: Selection and sorting with limited storage. Theoret. Comput. Sci. **12**(3), 315–323 (1980)
31. Muthukrishnan, S.: Data streams: Algorithms and applications. Now Publishers Inc. (2005)
32. Soviani, S., Kusnendar, J., Prabawa, H.: Learning how computer's work with combining CS-unplugged and raspberry Pi. In: Journal of Physics: Conference Series, vol. 1280, p. 032025 (2019)
33. Suo, X., Glebova, O., Liu, D., Lazar, A., Bein, D.: A survey of teaching PDC content in undergraduate curriculum. In: Proceedings of the 11th CCWC, pp. 1306–1312 (2021)
34. Szeider, S.: Large and parallel human sorting networks. In: Proceedings of the 7th CMSC, pp. 194–204 (2024)
35. Tonbuloğlu, B.: Tonbuloğlu, İ: The effect of unplugged coding activities on computational thinking skills of middle school students. Inform. Edu. **18**(2), 403–426 (2019)
36. Virtue, P.: GANs unplugged. In: Proceedings of the 35th AAAI, vol. 35, pp. 15664–15668 (2021)

Teaching Error Detection and Correction by Card Magic

Juraj Hromkovič and Regula Lacher(✉)

Department of Computer Science, ETH Zürich, Universitätstrasse 6, 8092 Zürich, Switzerland
{juraj.hromkovic,regula.lacher}@inf.ethz.ch

Abstract. In this paper we extend the activities from activity 4 "Card flip magic" of *Computer Science Unplugged*[1] We introduce several new card tricks for teaching error detection and error correction that enable the correction of up to three flipped bits. The main contribution is in designing a teaching sequence from kindergarten to high school, in which one starts with parity cards and finishes with famous self-detecting and self-correcting codes. Everything is designed in order to empower pupils to discover new codes with required properties to the greatest possible extent on their own.

Keywords: error detecting and correcting codes · computer science unplugged · genetic Socratic method

1 Introduction

We are living in the era of automation and AI. What consequences does it have for education? Automation has always been an integral part of human culture [1,2] with the goal of increasing our efficiency. Originally, people acquired knowledge and used it to develop procedures for reaching different goals. Before the computer era, the idea was that humans were able to learn to successfully apply these procedures without having the understanding of their inventors. Nowadays using IT we can automate any information processing that we can describe. The classical model of educational institutions striving to educate experts who are able to act correctly following a complex pattern needs to be essentially updated because acting by following a well-defined procedure is a thing of the past. Since education is about supporting humans to explore their (especially intellectual) potential, future education has to force the development of those dimensions of learners as creativity, fantasy and critical thinking, in which they can compete with information technology [3–5]. For this reason, our concept of teaching is based on our version of the following strategy called genetic Socratic method [6,7]:

Tim Bell, Ian H. Witten, Mike Fellows: Computer Science Unplugged. Off-line activities and games for all ages. (1998) https://classic.csunplugged.org/documents/books/english/unplugged-book-v1.pdf.

M. Löffler et al. (Eds.): CMSC 2026, LNCS 16951, pp. 27–40, 2027.
https://doi.org/10.1007/978-3-032-38887-2_3

"Do not focus on teaching the product of science and technology (facts, theorems, models, methods and equipment) and the use of them, teach the process of their investigation and development".

This is exactly the tradition of *Computer Science Unplugged* [8] that aims to empower learners to discover as much as possible on their own by active handling in the framework of well designed activities. This way of teaching corresponds closely to the constructivism of Jean Piaget [9, 10]. Our contribution in this paper follows exactly this path and shows how learners can step by step discover more and more advanced techniques for designing error-protecting codes including some well known codes. To reach this goal we start with magic cards from *Computer Science Unplugged* [8], add several new card magics and outline an activity moving from parity cards to self-correcting codes. More advanced tricks for university students are described by Greenberg [11].

Because one cannot completely avoid errors by entering or transmitting data, one of the basic possible requirements on information coding is that codes are robust (error resistant). As an error we consider flipping a bit in a binary code of data. If we use another alphabet (for instance the alphabet of all digits), then an error is based on exchanging one letter of the alphabet by another. Here we also consider errors when two consecutive symbols (letters) are exchanged in their order (for instance, 01 for 10, or AX for XA).

We distinguish three types of this robustness, for any $k \in \mathbb{N}$.

1. **k-error detecting codes.** If at most k errors happened, the receiver can recognize that something is wrong.
2. **k-unreadable correcting codes.** If at most k positions in the message are not readable, the receiver can reconstruct the original, missing content.
3. **k-error correcting codes.** The receiver can recognize up to k wrong symbols (errors) in the message and correct the content on the corresponding positions.

2 Card Magic and Parity Bits

In this chapter, we use card tricks to show the power of parity control bits, starting with the detection of one error and finishing with correcting three flipped bits. The complexity of card magic grows from kindergarten to high school.

We describe teaching through cards magic as a sequence of activities and creative workshops, similar to those in *Computer Science Unplugged*. Activities are guided by teachers, and the goal is to discover "how it works" (the functionality of computer science products). Based on the experience acquired in the activities, teachers ask pupils in creative workshops to discover and develop new concepts and products with desired properties, to verify the functionality of their own products, and to compare their products with the products of others. In this way, we support the exploration of the intellectual potential of the pupils and empower them to be creative and independent developers.

We use standard card sets in most of the activities and creative workshops. The exception is Activity 2, where we use specialized cards with beavers wearing three garments (cap, shirt, pants/skirt) of different colors (these cards are contained in [12]). This activity is very nice, as it is an application of Activity 1, but it is not essential for the overall learning objectives and can be omitted if the special cards are not available. Where diagonals are used for control cards, square cards are easier to work with. The sequence of activities and creative workshops should not be changed, as the activities build on the experience gained in the earlier ones.

Activity 1: One Parity Bit

Teacher's Guide: Note that, in addition to the teacher, a magician is required. This magician needs to be trained beforehand and can be another adult or some of the pupils. The teacher takes on the role of the helper (of the magician) (Fig. 1).

Fig. 1. How many cards have been flipped, one or two?

All pupils gather around a table, and the magician(s) leave(s) the room. Now the teacher asks for a volunteer among the pupils and this volunteer is asked to place some cards on the table, in a row, and some of the cards face-up (representing 1) and some other cards face-down (representing 0). The teacher gives cards out one by one to the volunteer, and the volunteer freely decides whether the cards are placed face-up or face-down. The teacher continues to give cards until an even number of face-up cards is on the table. The teacher thanks the first volunteer and asks for a second volunteer. This second volunteer is now instructed to flip one or two cards of his choice - if the card chosen was face-down, it is flipped to face-up and vice versa. Now the magician(s) are asked to enter the room and answer the question "How many cards have been flipped, one or two?" This is repeated several times until the pupils recognize how it works and can take on both roles, magician and helper.

Solution: Magician(s) calculate(s) the parity of the cards which are face-up and say "two" if this is even and "one" if it is odd. This works because any card being flipped changes the parity of the number of face-up cards, and the helper ensured that there was an even number of face-up cards before the flip(s).

Expectations and Extensions: The main educational benefit is that pupils discover that one control bit is sufficient to detect an error or to reconstruct one unreadable bit in an arbitrarily long sequence of bits. One can play this activity asking "zero or one" instead of "one or two", which more closely reflects the practical application of detecting whether there is an error or not. Alternatively, one can remove one card and the magician(s) are asked to say whether the removed card was face-up or face-down, i.e., to reconstruct the unreadable bit.

Activity 2: More Parity Bits and Reconstructing Unreadable Bits

Teacher's Guide: As in Activity 1, a magician is needed. A set of cards must be prepared beforehand. One color is selected as the main color (we use red as an example). A card set is then chosen such that each card shows only two colors (red and yellow in our example), and the total number of garments of the selected color (cap, shirt, pants/skirt) is even. All pupils gather around a table, and the magician leaves the room. The teacher asks one volunteer to arrange the prepared cards face-up on the table in an arbitrary order. The teacher then asks another volunteer to take any one of the cards and hide it (or turn it face-down on the table). The magician is invited back into the room and asked to determine the colors of the garments on the hidden card (Fig. 2).

Fig. 2. Set of beaver cards where the number of red garments (cap, shirt, pants/skirt) is even. (Color figure online)

Solution: The magician calculates the parity of the red caps among the visible cards. If the number is odd, the magician says "red cap" (because one red cap is missing); otherwise it is the other color (yellow in this example). The same strategy is applied to the shirt and pants/skirt.

Expectations and Extensions: Activity 2 is a straightforward extension of Activity 1, demonstrating how the same parity principle can be applied independently several times.

Since all cards are unique, students might mistakenly think that the magician determines the colors on the flipped card by the principle of exclusion rather than parity. One can use multiple copies of the same card to avoid this. Still better would be to combine it with the following extension. One can use an arbitrarily large set of beaver cards with garments of two colors distributed on the table. The magician can then demonstrate an extraordinary memory by observing the cards on the table and determining the colors of a removed card. This is possible because the magician needs to remember only three bits: the parities of the red garments in the original card set. The magician then compares these parities with those of the set of cards after one card is removed. Whenever the parity differs for one garment, that garment must be red on the removed card.

Activity 3: Correcting One Error
Since this activity is essentially the well-known Activity 4 of *Computer Science Unplugged*, we present it in the appendix.

After completing Activities 1, 2, and 3, pupils are asked to develop new card tricks that offer new methods for creating self-detecting codes.

Creative Workshop 1: Correcting One Error in the Triangle Layout

Teacher's Guide: The teacher arranges the cards (some face-up, some face-down) in the shape of a triangle, as shown in Fig. 3 a). The triangle can be of an arbitrary size n.

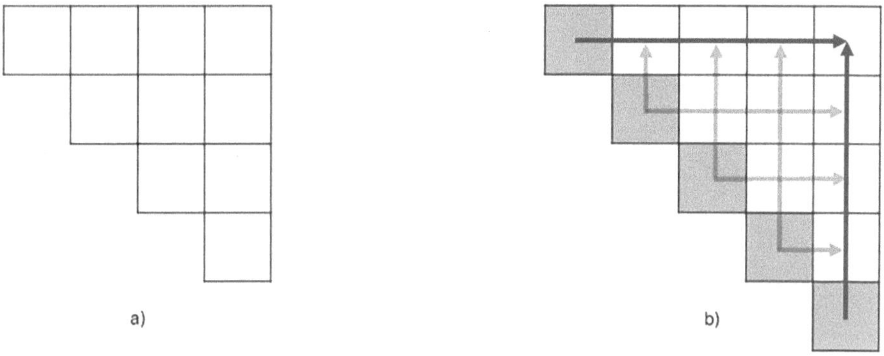

Fig. 3. a) Message bits (cards) in triangular layout. b) Every message bit (card) is monitored by two control bits (cards).

The teacher then asks the class to develop a new magic trick for correcting one error by using at most $n + 1$ parity control bits. The teacher encourages the class to experiment with different proposals, to test the functionality of each one, and to learn from unsuccessful attempts (productive failure).

Solution: One possible solution is to add $n+1$ control cards on the diagonal in such a way that the combined parity of the column above the control card and the row to the right is even. If a card in the original triangle is flipped, then two control cards on the diagonal indicate that an error has occurred. The flipped card is located at the intersection of the areas they monitor. If a control card is flipped, then this is the only one error that is detected and can be corrected.

Expectations and Extensions: Pupils learn that one can choose which area of message bits each control bit monitors. The important point is that each message bit is controlled by at least two different control bits.

Creative Workshop 2: Two Errors and the Original Magic Trick

Teacher's Guide: The teacher poses the question: Can we correct two flipped cards with the magic trick from *Computer Science Unplugged*, Activity 4? The best approach is to perform this card magic with two flipped cards, allowing the pupils to discover the correct answer and to explore all possible cases, as presented in the solution.

Solution: The answer is that this is not possible. Either there are two possible pairs of flipped cards (Fig. 4a), shown in blue and red), or, if the flipped cards are in the same row or column, we see six possible pairs of flipped cards (Fig. 4b) shown in six different colors).

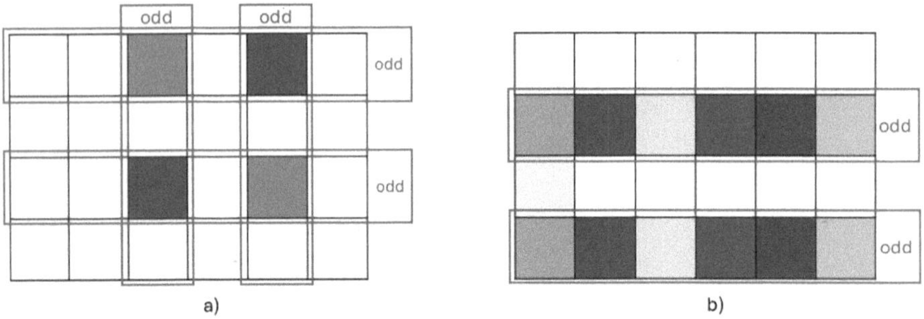

Fig. 4. Two flipped cards in the original *Computer Science Unplugged* card magic. a) The parity is incorrect in two rows and two columns. b) The parity is incorrect in two columns only.

Creative Workshop 3: Magic for Correcting Two Flipped Bits
Teacher's Guide: After establishing in Creative Workshop 2 that we cannot correct two errors using the classical magic trick, the teacher asks the pupils to try to develop a way of adding additional control cards that enables them to correct two errors. Again, the teacher encourages the pupils to develop different proposals and investigates all proposals by playing them with the cards. This is done with the whole class, and all classmates are asked to find a situation (two errors) for which the trick does not work. Pupils learn through this process of productive failure until they discover a working solution.

Solution: There is a natural solution based on the addition of control cards for the diagonals, as shown in Fig. 5. Using this strategy, one can easily recognize that error patterns such as in Fig. 4a) can be easily corrected because the additional diagonal control bits indicate which of the two possible error patterns has occurred. Likewise, in the case shown in Fig. 4b), the new control bits, together with the two column parity bits, unambiguously identify the positions of the flipped cards. Several cases must be distinguished when one or two of the new control bits are flipped.

Expectations and Extensions: Pupils recognize that two errors can be corrected if every message bit is monitored by at least three control bits. This card trick is somewhat more difficult to perform, but high school students can usually learn it quite quickly.

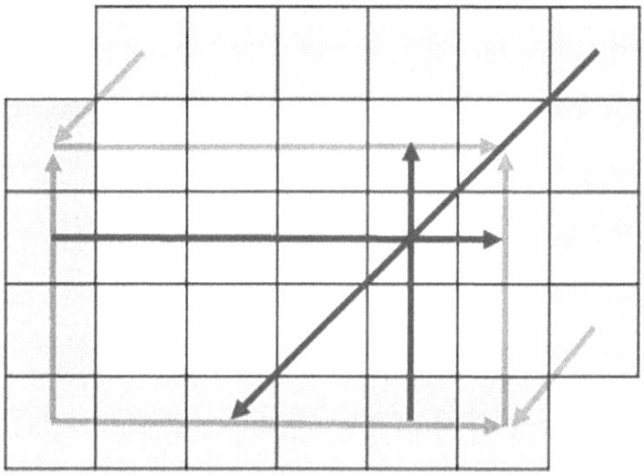

Fig. 5. Additional control bits (blue) monitor the diagonals from the top right to the bottom left as indicated by the arrows. Consequently, every message bit is monitored by three control bits, allowing the detection and correction of two errors. (Color figure online)

Creative Workshop 4: Magic for Correcting Three Flipped Bits

Material: A sufficiently large card set. Square cards are easier to use than rectangular cards.

Teacher's Guide: The goal is to extend the card trick from Creative Workshop 3 so that 3 errors can be corrected. The pupils first must discover for which error patterns this card trick no longer works. The teacher then suggests adding control cards (bits) to enable the correction of up to three errors. As usual, the pupils propose new card tricks, and the other pupils analyze whether these proposals are able to correct up to three errors, regardless of their positions.

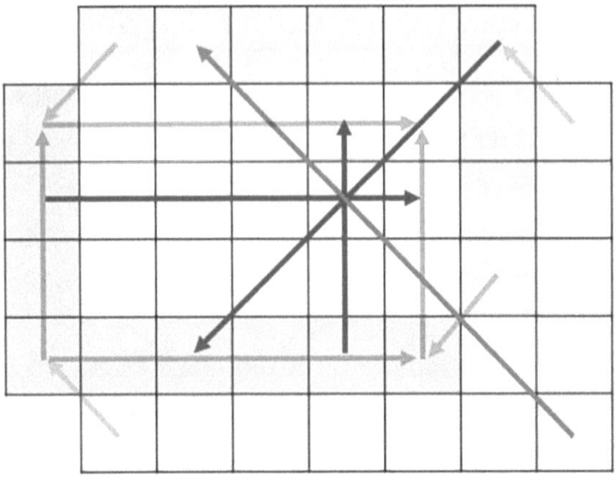

Fig. 6. Additional control bits (green) monitor the second set of diagonals, running from the bottom right to the top left. Consequently, every message bit is monitored by four control bits, allowing the detection and correction of three errors. (Color figure online)

Solution: The basic idea has already been introduced in the previous workshop. One adds control cards for a second group of diagonals. Figure 6 illustrates the construction. The red cards are the original control cards from Activity 3. The blue cards are the control cards introduced in Creative Workshop 3, which monitor the diagonals running from the bottom left to the top right. The green cards are the new control cards and they are monitoring the diagonals running from the bottom right to the top left.

Expectations and Extensions: After developing a two-error-self-correcting code in Creative Workshop 3, pupils will usually propose the strategy shown in Fig. 6. We cannot expect the class to prove its correct functionality when errors in the control bits are also allowed, because too many cases must be considered. Instead, the class can perform this card trick to explore different situations. For instance, one group takes the role of magician, while another group tries to flip at most three cards in such a way that the magician has a challenging task in detecting and correcting the three errors.

Creative Workshop 5: Detecting the Exchange of Two Consecutive Bits

Teacher's Guide: The goal is to improve the trick from Activity 1. In an arbitrarily long sequence of bits, we allow at most one error of the following types: either one bit is flipped, or the order of two consecutive bits is exchanged (i.e., "01" is exchanged for "10" or vice versa). Depending on the situation, the magician must distinguish between three possibilities: no error, one flipped bit, or the exchange of the order of two consecutive bits. The teacher asks the pupils to use a fixed number of control cards (control bits), regardless of the length of the sequence.

Solution: Two control bits are sufficient. One control bit monitors the parity of the bits in the odd positions of the binary sequence, and the second control bit monitors the parity of the bits in the even positions. After receiving a message containing one error, there are three possibilities. If the first control bit is incorrect and the second control bit is correct (the number 1 s in the odd positions is odd), then the flipped bit is in an odd position. If only the second control bit is incorrect, then the flipped bit is in an even position. If both control bits are incorrect, then the order of two different consecutive bits has been exchanged.

Expectations and Extensions: This creative workshop provides excellent preparation for introducing EAN codes because bits in even and odd positions are treated separately.

3 Magic with Digits

In the previous section, we presented a couple of card tricks based on control bits monitoring parity. The goal of this section is to extend these card tricks from bits to digits and use card magic to introduce the well-known EAN and ISBN codes.

The material used in this section are cards with digits $0, 1, ..., 9$ on their faces. In addition, Creative Workshop 7 uses a card with the value 10.

Activity 4: Digit Sum Modulo 10 for Error Detection

Teacher's Guide: Similarly to Activity 1, the teacher introduces a trick using digit cards. The magician leaves the room, and a volunteer arranges an arbitrary sequence of cards with values $0, 1, ..., 9$, for example 03544. The teacher calculates the sum of the digits shown on the cards. In our example, $0+3+5+4+4 = 16$. The teacher adds one control card with value 4. In general, the value of the control card is chosen so that the sum of all digits of the sequence is a multiple of 10 (the sum of digits $modulo 10$ is 0). Afterwards, another volunteer hides (flips) one card. The magician is invited back into the room and asked to determine the value of the hidden card.

Solution: The magician computes the sum of the digits on the visible cards. The value of the hidden card is the number that must be added to this sum to obtain the next multiple of 10.

Expectations and Extensions: After performing this card trick several times, the pupils discover the underlying principle, building on their experience with the parity-based card tricks from Section 2. An extension is to ask the second volunteer either to remove one card or to leave all cards unchanged. The magician must then answer "0" or "1", depending on whether no card or one card has been removed. Another variation is to use only cards with the values $0, 1, 2$, and to choose the control card so that the digit sum is a multiple of 3.

Activity 5: One-Error Correction with Digit Cards

Teacher's Guide: This activity is a natural extension of the card trick from Activity 3. One volunteer arranges the digit cards face-up in a rectangle. The teacher then adds one row and one column with control cards in such a way that the sum of digits in each row and each column is a multiple of 10. Another volunteer replaces one card in the rectangle (including the control cards added by the teacher) with a different card. The magician is invited back into the room and asked to identify the incorrect card and replace it with the correct one.

Solution: The magician calculates the digit sums of all rows and all columns. The magician identifies the row and the column whose sums are not multiples of 10. The incorrect card is located at the intersection of this row and column, and the magician can determine the correct value of the replaced card.

Expectations and Extensions: Using the experience gained from Activities 3 and 4, the pupils can quickly discover the principle behind this card trick. The activity can also be played with a smaller set of digit values, for instance $0, 1, 2, 3, 4, 5$ and then counting modulo 6. Again, the pupils can discover why this card trick does not work for correcting two errors and investigate how additional control cards can be introduced to enable the correction of two errors.

Activity 6: Detection of One Error, Including an Exchanged Order

Teacher's Guide: Similar to Activity 4, a volunteer places some digit cards face-up, for instance 2 7 1 2 3. The teacher then adds two control cards, resulting in the sequence 2 7 1 2 3 4 1. The value of the first control card guarantees that the sum of the digits in the *odd* positions is a multiple of 10, and the second control card guarantees that the sum of digits in the *even* positions is a multiple of 10. Another volunteer is then asked to do one of the following three actions: do nothing, replace one card with another, or exchange the order of two consecutive cards. Finally, the magician determines which of the three actions has been performed.

Solution: If both control cards are correct, the magician knows that nothing has been changed (or more than one error has occurred). If both control cards are incorrect, then the order of two consecutive cards with different values has been exchanged. If exactly one of the control cards is incorrect, then one card has been replaced by another.

Expectations and Extensions: After performing this card trick a few times, pupils will discover the underlying principle. One can use this idea to extend the card trick from Activity 5 so that it can also correct a switched order of cards in the rectangle, but this is a more advanced challenge.

Creative Workshop 6: EAN Codes

Teacher's Guide: EAN codes are not a result of an optimization process but rather of a suboptimal consensus. Because of this, it is not appropriate to ask pupils to develop them by specifying a set of requirements. Instead, we suggest introducing EAN codes and asking the pupils to investigate their properties. EAN codes use only one control digit, and can detect one error when one digit is replaced by another. In approximately $80-90\%$ of cases, EAN codes can detect switches of the order of two consecutive values.

Suppose we have the digit sequence 27123. We compute the value of the control digit as follows. We add the values of the odd positions and three times the values in the even positions:

$2+1+3+3\cdot(7+2)=6+27=33$

The control digit is 7 because $33+7=40$ is the next multiple of 10. After checking several sequences to determine whether they are valid EAN codes, the teacher can pose the following tasks:

(i) Can an EAN code always detect one error (replacing one digit with another)?
(ii) Can an EAN code reconstruct one unreadable digit?
(iii) Can an EAN code detect an error when two consecutive digits are exchanged? If not, find examples when this can happen.

Solution: Explaining the answers to these questions requires too much space in this paper. Therefore, we refer to Chap. 4 of *Data Science and Sicherheit* in [13].

Creative Workshop 7: ISBN Codes

Teacher's Guide: ISBN codes also use only one control digit (which may additionally take the value 10). In contrast to EAN codes, ISBN codes can detect not only a single erroneous digit but also the transposition of two consecutive digits. For a sequence $x_1, x_2, ..., x_9$ of nine digits, the control digit x_{10} is calculated so that $10\cdot x_1+9\cdot x_2+8\cdot x_3+7\cdot x_4+6\cdot x_5+5\cdot x_6+4\cdot x_7+3\cdot x_8+2\cdot x_9+x_{10}$ is a multiple of 11. The modulus 11 is chosen because it is a prime, which implies that x_{10} could also have the value of 10.

After the pupils have learned to distinguish valid ISBN codes from invalid ones, the teacher can pose the following tasks.

(i) Can an ISBN code always detect the replacement of one digit by another? Justify your answer.
(ii) Can an ISBN code always detect the transposition of two consecutive digits? Justify your answer.
(iii) Can an ISBN code detect the transposition of two digits in arbitrary positions within the sequence?

Solution: Since answering these questions requires more space than is available in this paper, we refer to Chap. 4 of *Data Science and Sicherheit* in [14].

Expectations and Extensions: There are many opportunities to deepen the understanding of this code. For example, would it still work if we used modulus of 10 instead of 11? Would it work if we used only the digits $0, 1, 2, 3, 4, 5, 6$ and performed all calculations modulo 7?

4 Conclusion

In this paper, we designed a sequence of lessons for teaching error detection and error correction through card magic. In this way, we have extended the corresponding chapter of *Computer Science Unplugged* [8]. The added value of our contribution lies in two dimensions. First, our sequence of card tricks enables the teaching of this topic in a spiral curriculum from kindergarten to high school and beyond. Second, our approach actively involves pupils in the design of error-robust codes. In this way, pupils do not only understand how these codes work, but are also empowered to develop codes with required properties to a large extent on their own. Based on the sequence presented here, pupils can develop Hamming codes without using any mathematical concepts beyond those introduced in this paper. Explaining how to derive Hamming codes from these card tricks would require more space than is available here, however, it is presented in [15] and [16].

Acknowledgment. We thank Vaidotas Kinçius for designing the drawing of the beaver cards.

Disclosure of Interests. The authors have no competing interests to declare that are relevant to the content of this article.

A Activity 3: Correcting One Error

Activity 3 has been moved to the appendix because it is essentially the well-known activity 4 from *Computer Science Unplugged* (Fig. 7).

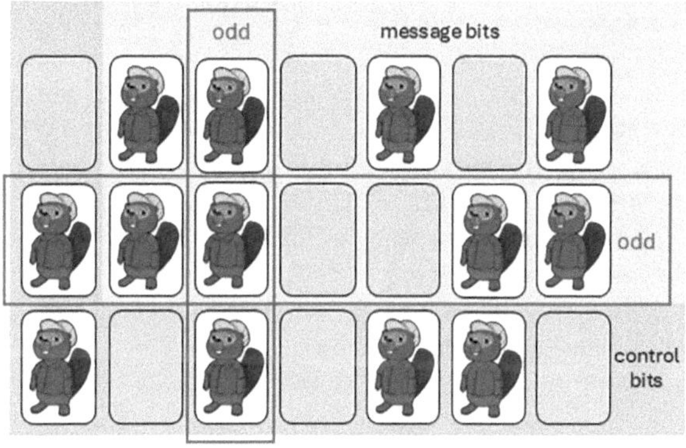

Fig. 7. The error is located at the intersection of the areas monitored by those control bits for which the number of face-up cards is odd.

Teacher's Guide: This is the well-known Activity 4 from *Computer Science Unplugged* [8], in which a volunteer arranges a rectangle of face-up and face-down cards, and the magician's helper extends the rectangle by one row and one column so that the number of face-up cards in every row and every column is even.

Solution: If one card is flipped, then the parity of the corresponding row and column becomes odd, and the flipped card is located at the intersection of that row and column.

New Extension: Instead of flipping one card, one can remove two cards (i.e., make two cards unreadable). Pupils are asked to determine whether the removed cards were face-up or face-down. This is possible because the cards must be reconstructed so that the parity of face-up cards in all rows and columns becomes even.

References

1. Hromkovič, J.: Homo informaticus. why computer science fundamentals are an unavoidable part of human culture and how to teach them. In: Bulletin of EATCS, vol. 115, pp. 112–122 (2015). http://hdl.handle.net/20.500.11850/110737
2. Hromkovič, J., Lacher, R.: The computer science way of thinking in human history and consequences for the design of computer science curricula. In: Dagiene, V., Hellas, A. (eds.) ISSEP 2017. LNCS, vol. 10696, pp. 3–11. Springer, Cham (2017). https://doi.org/10.1007/978-3-319-71483-7_1
3. Dagienė, V., Hromkovič, J., Lacher, R.: Designing informatics curriculum for K-12 education: from concepts to implementations. Inf. Educ. 20(3), 333–360 (2021). https://doi.org/10.15388/infedu.2021.22

4. Dagiene, V., Hromkovic, J., Lacher, R.: A two-dimensional classification model for the bebras tasks on informatics based simultaneously on subfields and competencies. In: Kori, K., Laanpere, M. (eds.) ISSEP 2020. LNCS, vol. 12518, pp. 42–54. Springer, Cham (2020). https://doi.org/10.1007/978-3-030-63212-0_4
5. Hromkovič, J., Lacher, R.: How teaching informatics can contribute to improving education in general. In: Bulletin of EATCS, vol. 139. (2023). https://bulletin.eatcs.org/index.php/beatcs/article/view/752/794
6. Hromkovič, J., Lacher, R.: Teaching design of algorithms in high schools by constructive induction. In: Bulletin of EATCS, vol. 144 (2025). http://eatcs.org/beatcs/index.php/beatcs/article/view/821
7. Wagenschein, M.: Zum Problem des Genetischen Lehrens. Zeitschrift für Pädagogik (1966)
8. Bell, T., Witten, I. H., Fellows, M.: Computer Science Unplugged. Off-line activities and games for all ages (1998). https://classic.csunplugged.org/documents/books/english/unplugged-book-v1.pdf
9. Piaget, J.: La représentation du monde chez l'enfant (1926)
10. Piaget, J., Harel, I.: The Psychology of Intelligence. Harvard University Press, Cambridge (1950)
11. Greenberg, R.: Educational magic tricks based on error-detection schemes. In: ITiCSE '17, Bologna, Italy, 03–05 July 2017. ACM (2017). 978-1-4503-4704-4/17/07. https://doi.org/10.1145/3059009.3059034
12. Hauser, U.: Rätsel und Spiele ohne Computer, Karten (in German). einfach Informatik, Zyklus 1. Handbuch, Klett und Balmer AG, Baar, Switzerland (2020). ISBN 978-3-264-84809-0
13. Barot, M., et al.: Data Science und Sicherheit. In: German. Part of a series of textbooks for high schools, Klett und Balmer AG, Baar, Switzerland (2022). ISBN:978-3-264-84597-6
14. Barot, M., et al.: Data Science und Sicherheit. In: German. Part of a series of textbooks for high schools, Klett und Balmer AG, Baar, Switzerland (2022) (2020). ISBN: 978-3-264-84597-6
15. Hamming, R. W.: Error detecting and error correcting codes. Bell Syst. Techn. J. **29**(2), 147–160 (1950). https://doi.org/10.1002/j.1538-7305.1950.tb00463.x
16. Hromkovič, J., Lacher, R.: Teaching to think in informatics lessons by genetic socratic method illustrated by discovering hamming codes. In: Invited talk at ISSEP 2026. LNCS. Springer, Cham (2026)

Hangers and Straws: Designing Low-Cost Unplugged Activities for Primary-School Algorithms

Michael Lodi[1,2(✉)]

[1] Department of Computer Science and Engineering, University of Bologna, Bologna, Italy
michael.lodi@unibo.it
[2] Laboratorio CINI Informatica e Scuola, Rome, Italy

Abstract. Recent Italian curriculum guidelines explicitly introduce informatics in primary school, distinguishing it from generic digital competence and emphasising data, algorithms, and programming. This paper describes the design rationale for two low-cost unplugged activities for the first years of primary school. In the first activity, learners use a clothes hanger as a balance scale to find the heaviest opaque bag. The available actions are restricted to a small instruction set, so that the natural solution exposes the invariant of the current champion. In the second activity, learners order five coloured straws while an adult helper can only compare two selected straws and swap them if they are out of order. A covered version hides the lengths and turns sorting into a local compare-and-swap problem. Both activities are accompanied by lightweight online companions, intended as plugged interpreters of the same rules. The contribution is a design analysis grounded in informatics education literature: explicit executors, computationally authentic constraints, discovery of strategies before naming algorithms, reflection on correctness and efficiency, and careful transition from unplugged to plugged work.

Keywords: Informatics education · Primary school · Algorithms · CS Unplugged · Sorting · Maximum search

1 Introduction

Sorting a handful of objects and finding the largest among them are among the oldest and most recognisable problems in computing, and they have a long history as accessible, hands-on activities for young learners [1,4]. They are appealing precisely because they can be posed with everyday objects and a single, easily stated rule, yet they open onto deep ideas: the existence of a precise procedure for an *executor*, the notion of an *invariant*, the distinction between a problem and an algorithm that solves it, and the idea that algorithms can be compared by counting the elementary operations they perform.

M. Löffler et al. (Eds.): CMSC 2026, LNCS 16951, pp. 41–55, 2027.
https://doi.org/10.1007/978-3-032-38887-2_4

This paper describes the design of two unplugged activities—and of two web simulators that act as their "plugged" companions—that we created for a commercial primary-school product and an openly licensed teacher guide. Our aim is to make explicit the design decisions and the research and disciplinary principles behind them. We pay particular attention to one principle that, in our experience, is easy to lose when informatics enters the primary curriculum: that—in this discipline—an algorithm can be seen as a precise, unambiguous procedure *with respect to a specific executor*, and not merely an informal "recipe" [8,33]. Both activities are organised around making that executor visible and rigorous, even without a computer.

Methodologically, this paper is a curriculum design report informed by design research principles. It documents the rationale of a set of artefacts before their first systematic field cycle, so that the design itself—not yet its effects—can be examined and criticised. The commission came with concrete and tight constraints: the activities had to work in any Italian classroom, from the first primary years onwards, with non-specialist teachers, materials of negligible cost, and the page format of a student booklet, while remaining computationally authentic. We condense these constraints into two design questions: (1) how can classic information-processing problems (maximum finding, sorting) be staged physically so that the executor is explicit and the constraints are embodied in the material situation—in the objects, the roles, and what can be perceived—rather than left as rules for the teacher to enforce? (2) how can a digital companion extend the activity (scale, repetition, feedback) without changing the executor children have learned to control?

Section 2 recalls the curricular and research background; Sect. 3 distils it into five design principles; Sect. 4 presents the two activities and their companions as applications of those principles; Sect. 5 connects them to expected learning outcomes and assessment opportunities.

2 Background and Motivation

2.1 Informatics in the New Italian Curriculum

The 2025 revision of the Italian *Indicazioni Nazionali* (national curriculum guidelines) for the first cycle of education (primary school, ages 6–11, and lower secondary school, ages 11–14) introduces, for the first time, explicit references to informatics (albeit within the disciplines *Mathematics* and *Technology*) starting from primary school [32]. An extended analysis of the informatics content of the new guidelines can be found in [28].

The document explicitly distinguishes digital competence, understood as responsible and effective use of digital devices and services, from informatics competence, understood as the comprehension of how data can be automatically processed by agents that mechanically execute human-designed procedures [32]. In the primary-school mathematics section, the expected competencies include representing information as data, describing practical tasks through algorithms, writing and understanding simple programs in educational languages, and evaluating whether they are adequate for the task one wants to automate [32].

This curriculum shift creates a practical need for learning materials that are scientifically grounded but usable by primary teachers. Analyses of current textbooks show that, where computing already appears, it is most often presented as transversal "coding"—a term used, especially in Italy, with an excessively broad meaning [13,29]. Many proposals are sound and engaging, but a recurring weakness is that activities are presented without a genuine, well-defined interpreter: the link between an algorithm and the *automatic processing of information* it is supposed to describe is left implicit, and the very notion of "algorithm" is often conveyed through coarse metaphors such as the "cooking recipe" [33]. As Monga and colleagues argue, such metaphors neglect a key idea that predates electronic computers: to define an algorithm unambiguously—and thus assess its effectiveness—one must make explicit an *interpreter*, a computational agent able to carry out certain elementary operations and not others [30,33]. Studies of how students and teachers conceptualise "algorithm" confirm that this executor-relative character is often missing from informal accounts [8].

It was against this background that a major Italian publisher (Mondadori Education) asked the author and other Italian colleagues to introduce informatics content into a primary-school product. The activities discussed here (both the unplugged versions and their online companions) were designed by the author as part of an informatics booklet for the primary-school textbook *Sulle orme di Milù* [10]. Most prominently, a detailed teacher guide, specific to informatics, accompanies the booklet [9]; a derived version of the guide is published under a Creative Commons licence [27][1]. The editorial process behind the booklet and the open guide is described in [26].

2.2 Unplugged Approaches: Promises and Limits

Unplugged activities (teaching informatics without a computer) have become a popular way to introduce fundamental concepts [1,4]. They offer a constructivist setting in which, by manipulating real objects, learners can observe, hypothesise, and validate; they require inexpensive materials; and they remove the technological hurdles of having to acquire programming fluency before engaging with ideas such as algorithms [3]. The official CS Unplugged principles emphasise, for unplugged activities, no computers required, real informatics, learning by doing, fun, no specialised equipment, cooperative work, and resilience to errors [14].

Research, however, gives a balanced picture. CS Unplugged can provide access to deep ideas and support teachers, but unplugged work alone does not guarantee that students connect the activity to informatics concepts [5]; studies have reported limited changes in attitudes or perceived content understanding when this connection is weak [18,37]. Conversely, combined unplugged and plugged approaches can be productive when the bridge is explicitly designed:

[1] https://insegnareinformatica.github.io.

Hermans and Aivaloglou found benefits of an unplugged-first approach in a controlled comparison of Scratch programming lessons [22], and Bell and Lodi argue that unplugged activities work best as part of a "both... and" approach, with reflection phases and plugged follow-ups that keep the computation in view [2,3]. These findings directly shape principles P2 and P5 below.

2.3 A Lineage: From Balance Scales to a Clothes Hanger and Straws

Both activities descend more or less directly from the CS Unplugged sorting activity "Lightest and Heaviest" [7], in which children sort weights using balance scales that compare two items at a time, count the comparisons, and even derive that selection sort costs $1 + 2 + \cdots + (n - 1) = n(n - 1)/2$ comparisons. That activity was adapted in Italy with purpose-built balance scales by Davoli and colleagues in Bologna, within primary-school dissemination projects [6,24,25].

The two activities presented here—described in full in Sect. 4—continue this tradition: in the first, children find the heaviest of four opaque bags using a clothes hanger as a balance scale; in the second, they sort five coloured straws by pointing at two of them, while an adult helper swaps them only if they are out of order. We narrow and sharpen that lineage for the first primary grades: the clothes-hanger activity isolates *maximum finding* (a single scan with a champion invariant), while the straw activity reduces the executor to a *single* compare-and-swap primitive and adds a "covered" phase that forces reasoning from feedback.

The straw activity is also directly inspired by the developmental work of Yang, Thompson, and Kidd [38]. They found that children aged four to ten performed above chance on a task in which object heights were hidden and could be inferred only through pairwise comparisons, and that many children spontaneously used behaviours identifiable as selection sort or shaker sort.

3 Design Principles

We distil the background above into five principles. They are stated here once, and Sect. 4 shows how each activity realises them.

P1—The Executor Comes First. The notion of "algorithm", and its didactical implications for the first cycle, are actively debated [8,16]; here we adopt the characterisation used by the new national guidelines, where an algorithm is a precise, unambiguous description of a procedure *with reference to a specific executor* [30,32,33]. Without that executor, the questions of unambiguity, termination, and effectiveness cannot even be posed. Each activity is therefore built around an explicit *instructions manual*[2]: a small, closed set of elementary operations—here in a physical game rather than software. An explicit executor also lets us distinguish a *computational problem* (find the heaviest of *any* set of

[2] The name, borrowed from appliance manuals, is deliberate: it does not tell learners what to do—that is the program's job—but lists what the executor *is able* to do, its entire repertoire of elementary operations.

bags; sort *any* sequence of straws) from a single instance [15], and ask whether a procedure is correct for all instances and at what cost—ideas usually missing from the "recipe" metaphor [8, 33].

P2—Constraints Live in the Situation, Not in the Rule Book. In many classroom games, restrictions are conventions that the teacher must state and police. Here, instead, the constraints are embodied in the material situation, in the spirit of Brousseau's *milieu*, whose feedback comes from the objects rather than from the teacher [12]. Information is physically degraded—bags are opaque, straw lengths are covered—so that the primitive operations are the *only* access to the state of the game, and the props and roles (a hanger that holds at most two bags, a helper who alone may touch the straws) make illegal moves awkward rather than merely forbidden. Under hidden information, a comparison can also be seen as an *epistemic action* in the sense of Kirsh and Maglio: an action whose value lies in the information it yields, not only in the change it produces [23]. Local, feedback-driven strategies are then not a rule to obey but the only rational option, and the constraint itself carries the informatics content: reasoning under partial information obtained through well-defined queries.

P3—Invention Before Naming, with Guidance. Standard topics can become catalogues of names. For young children we deliberately aim at the *invention* of a strategy, not the memorisation of a named algorithm: Futschek argues that algorithmic thinking is best developed by having learners invent and play algorithms themselves, independently of any programming language [19], and the evidence that children discover efficient sorting strategies on their own, even under hidden information [38], reassures us that the tasks are age-appropriate. This is not an argument for unguided exploration: pure discovery is ineffective, while *guided* discovery with careful scaffolding is effective [31]—here the scaffolding is the instructions manual and the staged phases, which let children construct knowledge while steering them to the goal [25]. By contrast, older students too often memorise the sorting algorithms that are prominent in textbooks without understanding them: misconceptions about basic algorithms, their invariants, and their efficiency have been shown to persist even after traditional instruction, up to university level [17, 20]. Starting from invention, with the executor in view, is meant to forestall that.

P4—Everyday Materials at Negligible Cost. A classroom balance scale is not always available, and plastic ones are often expensive and clunky, or imprecise for close weights. While teachers can build their own [25], this is not always possible, especially if activities are to be proposed nationwide. A clothes hanger is a familiar object that can be found at negligible cost and is fun to use for a purpose radically different from its original one; paper straws eliminate the comparison device altogether, replacing it with an adult helper. In designing the activities we adhered to the CS Unplugged design pattern of Nishida and colleagues[3]: choose an informatics concept, identify its key elements, map them to common objects or games, turn the situation into a challenge, evaluate, refine, and publish with guidance for teachers [34].

P5—Plugged Companions at the Same Abstraction Level. When moving from unplugged to plugged, a tempting but, in our view, mistaken step is to ask children to *re-implement* the activity in a general-purpose language—for instance, to build the sorting game in Scratch. For young learners this may be too hard, and, more fundamentally, it changes the problem: re-implementation means modelling data structures, input and output, and the comparison oracle itself, which shifts the cognitive load from the target concept to the accidental complexity of the language [11,36]. The constructive alternative is a high-level interpreter whose primitives are *matched to the activity*, consistent with learner-centred design arguments that programming tools should fit the learner's goals rather than maximise technical authenticity at all costs [21,35]. Companions should also respect the unplugged spirit on the infrastructural side, and be introduced *after* the physical activity, so that children connect the two experiences rather than treat one as a replacement for the other.

4 The Two Activities

The two activities share a common structure. The teacher introduces a physical problem using everyday materials (P4), makes the available operations explicit (P1, P2), and asks learners to invent a strategy rather than execute a named algorithm (P3). A later discussion connects the strategy to informatics concepts. The activity can then be revisited through a digital simulator that enforces the same rules (P5).

4.1 Hanger Balance: Finding the Heaviest Bag

The first activity asks learners to find the heaviest of four opaque bags. The bags look as similar as possible but contain objects with different weights (P2). A clothes hanger with two clips acts as a simple balance scale (P4). Two boxes or delimited areas are labelled "bags to weigh" and "eliminated bags". At the beginning, all bags are in the first box; at the end, exactly one bag, the heaviest, must remain on the hanger.

The key design element is the restricted instructions manual (i.e., the *instruction set*) prescribed by P1. The learner-programmer can use only four actions: take one bag from the "to weigh" box into the hand; hang the bag in the hand on the hanger; take the lighter of the two bags on the hanger into the hand; put the bag in the hand into the "eliminated" box. A further rule states that at any moment the learner can hold only one bag. The manual thus defines a tiny

[3] The evaluation step has so far been limited to informal feedback; systematic classroom evaluation is planned.

executor with one comparison primitive and a fixed repertoire of moves, mirroring the classic constraint in comparison-based algorithms. Figure 1 summarises the resulting physical machine.

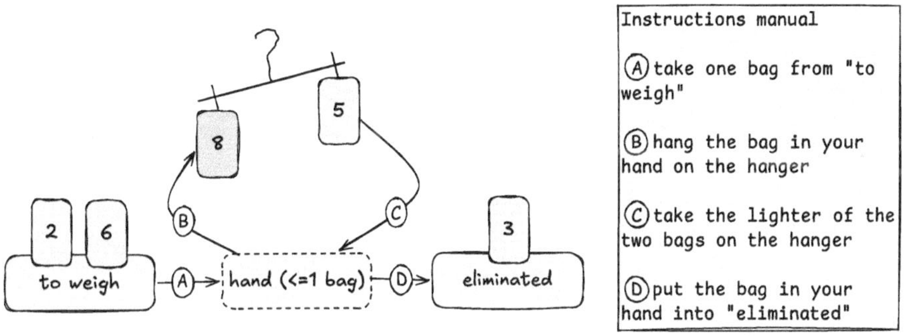

Fig. 1. The clothes-hanger activity as a physical machine. Left: the state (two boxes, the hanger, one hand holding at most one bag); the arrows are the only legal moves, labelled as in the instructions manual (right). Weights are hidden from the player and shown only to explain the invariant.

The natural strategy is to put two bags on the hanger, eliminate the lighter, bring in a new challenger against the survivor, and repeat. This is maximum finding by linear scan, and its correctness rests on an *invariant*: the bag left on the hanger is always the heaviest seen so far. In their worksheet, students are required to find the fully explicit fourteen-step program for four bags.

The tedium of writing out the steps for a larger number of bags lets the teacher motivate the introduction of repetition: first establish a champion, then repeatedly challenge it with a new bag until no unprocessed bags remain. The teacher guide stages this passage through three notations for the same strategy: the fully explicit sequence, an intermediate version with a jump ("if there are still bags to weigh, go back to instruction 3"), and finally a `while` loop—a miniature of the historical move from jumps to structured control [27].

The activity opens several discussions without requiring formal notation. Why is the eliminated bag safe to discard? How many bags can still be the heaviest after the first comparison? What happens if the same reasoning is repeated three times? What must the executor do if two bags have exactly the same weight? The last question is important because it reveals that an algorithm must specify behaviour for all relevant cases: if equal weights are allowed and the instruction says "take the lighter bag", the executor is under-specified.

The problem is also a low-threshold introduction to optimality. Finding the maximum of n distinct hidden weights by pairwise comparison requires at least $n-1$ comparisons, because every non-maximum element must lose at least once. The champion strategy achieves this bound. The point for early primary learners is, of course, not a formal lower-bound proof, but the physical situation can support the intuition: a bag that has never been compared and lost cannot yet be eliminated.

Following P5, a static, open-source online companion is available at https://lodi.ml/gruccia[4]. The companion offers a free exploration mode and a program mode. In free mode, users experience a simulation of the exact same unplugged activity. In programmer mode (Fig. 2), the blocks the child can assemble are *exactly* the four instructions of the manual, plus the `while` loop; the program is then run by a faithful interpreter on the simulated scene. The number of bags can be raised from four to eight, and a hint makes the cost of explicitness tangible (e.g. "with six bags an explicit solution needs 22 blocks, or 7 with the loop"), motivating abstraction through iteration.

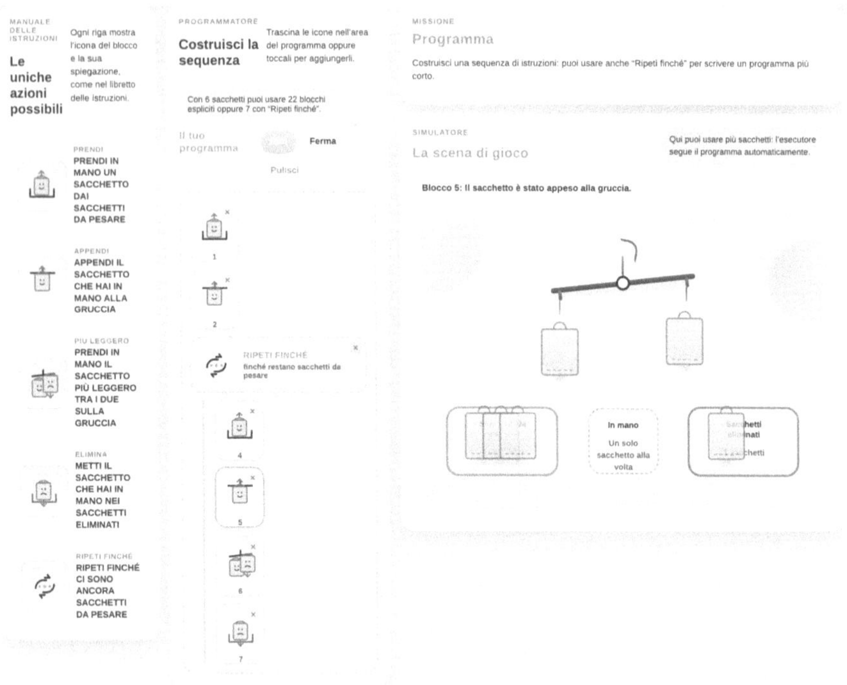

Fig. 2. Programmer mode of the clothes-hanger simulator (in Italian), captured mid-execution with six bags. Left: the palette *is* the instructions manual of Fig. 1, plus the `while` loop (*ripeti finché*). Centre: the seven-block champion program; the interpreter is executing block 5 (highlighted) inside the loop, and a hint recalls that an explicit solution would need 22 blocks. Right: the simulated scene mirrors the physical state—two bags on the tilted hanger, one already eliminated, and the one-bag hand.

[4] Source code at https://github.com/michaellodi/gruccia.

4.2 Covered Straws: Sorting by Compare-and-Swap

The second activity asks learners to order five coloured paper straws (P4) from shortest to longest. The activity is framed within a playful story about a "tidy barista" who is unable to sort the straws.

Here, the instruction set prescribed by P1 has a *single* instruction that the children must understand thoroughly: the child points at two straws; if the left one is longer than the right one, an adult helper swaps them; otherwise, the helper does nothing. The elementary operation is therefore a *compare-and-swap*: a comparison whose only observable effect is a possible exchange (Fig. 3). The adult, not a child, plays the executor, both to avoid left/right confusion and because, in the second phase, the helper must keep the straws covered while swapping them.

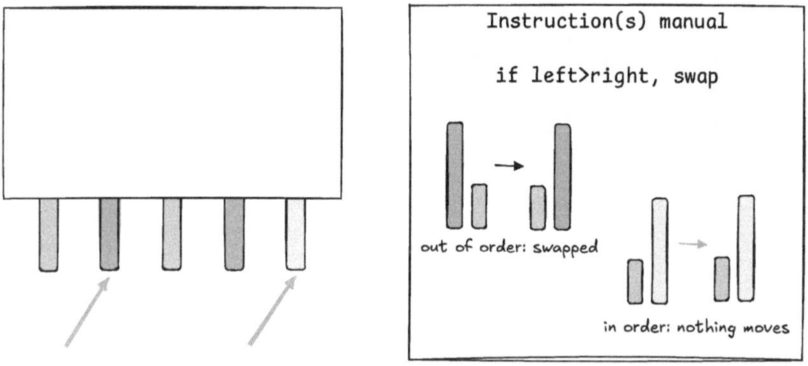

Fig. 3. The covered-straw activity. Left: a sheet covers the straws, so the learner sees positions and colours but not lengths, and selects two positions (arrows). Right: the executor's entire instructions manual—a single compare-and-swap—and its two possible outcomes: the state either changes or does not, and both results are informative.

The activity has two phases. In the uncovered phase, all lengths are visible. Learners can use global visual information, for instance, by finding the shortest straw and moving it toward the left. This phase is useful because it lets them formulate strategies and experience the compare-and-swap rule without an excessive memory burden. In the covered phase, the lengths are hidden from the student (but visible to the adult helper) behind a sheet of paper; only the colours and positions remain visible. Following P2, direct global inspection is no longer possible: the *only* information comes from the compare-and-swap feedback, and children must adopt a *local* strategy, trying an exchange, observing its effect, and deciding step by step how to continue.

Several strategies can emerge (P3). One family compares adjacent pairs from left to right, repeating passes until a pass makes no swaps. If learners alternate directions, the behaviour resembles *shaker sort*; if they move only left to right, it resembles *bubble sort*. Another family fixes a position and compares the straw in that position with later positions, swapping immediately when a shorter straw is found. This is related to *selection sort*, but with immediate exchanges, since the only primitive combines comparison and a possible swap. Of course, these names should not be presented to students: at this level, the goal is for learners to invent, describe, test, and refine strategies; naming can come much later as a way to connect their discoveries to known algorithms. The covered phase is intentionally close to the developmental work on hidden-rank sorting (see Sect. 2.3).

Important reflection questions can be raised in this activity as well. For example, one can ask children how to determine whether different strategies are more or less efficient (e.g., "by counting the number of comparisons"). Later, students can also be asked to justify why a strategy is correct in all cases (e.g., "the longest straw is in the last position after the first pass").

Following P5, a static, open-source online companion is available at https://lodi.ml/cannucce[5]. It offers an *uncovered* game and a *covered* game with the lengths hidden behind an on-screen sheet (Fig. 4); it counts comparisons and swaps, and lets the user choose the initial arrangement of the straws (random, nearly sorted, reversed). These features turn the aforementioned reflection questions into something children can investigate more quickly, running the same strategy on different inputs and counting comparisons, or discovering new strategies when an adult helper is unavailable.

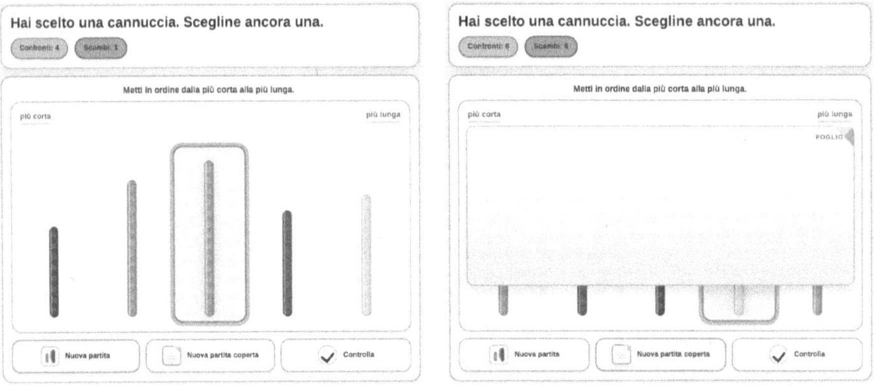

Fig. 4. The straw-sorting simulator (in Italian), with one straw selected in both games. Left: *uncovered* game—lengths are visible, so global visual strategies remain possible. Right: *covered* game—an on-screen sheet (*foglio*) hides the lengths, only colours and positions remain visible, and the compare-and-swap feedback becomes the only source of information. Counters track *comparisons* (*Confronti*) and *swaps* (*Scambi*); the initial arrangement can be set to random, nearly sorted, or reversed.

[5] Source code at https://github.com/michaellodi/cannucce.

4.3 How the Companions Instantiate the Bridge

The simulators instantiate the unplugged–plugged bridge of P5 in a deliberately constrained way: instead of an open programming environment, each offers an interpreter at the level of the activity's primitives—in the clothes-hanger programmer mode, the blocks *are* the four manual instructions plus a loop (Fig. 2); the straw simulator exposes only the compare-and-swap primitive. Children thus confront a real, rigorous interpreter and a real formalism, at a very high level, while repeated trials become cheap and fast.

The companions also respect the unplugged spirit on the infrastructural side: no build step, no server, no account, and no external dependency is required. They are meant to be introduced *after* the unplugged activity, so that children connect the two experiences. Moreover, their multimodal, interactive nature may support inclusive practice for students with special needs, and they are useful when the activity cannot be performed in person (e.g., online teaching, or large cohorts of students or teachers-in-training).

5 Expected Learning Outcomes and Assessment Opportunities

The activities operationalise specific expectations of the new guidelines [32]. Among the end-of-primary competencies in mathematics, they target "describing procedures for practical tasks through algorithms"—moving, however, beyond everyday tasks to genuinely computational ones (finding a maximum, sorting)—and "writing and understanding simple programs in educational languages, and evaluating their adequacy": each instructions manual is precisely such a language, small enough to be mastered in a single lesson. Among the objectives for the first three grades, they address describing tasks as precise, unambiguous sequences, and verifying programs by executing them—with the peculiarity that the executor is a person, so execution, verification, and debugging happen in the open. In terms of observable outcomes, after the activities and the guided reflections learners should be able to: produce a *legal* program (only instructions from the manual, never leaving the executor stuck); predict the effect of an instruction on the state of the game; apply a strategy unchanged to new instances (new weights, new lengths), separating the problem from the single instance; justify informally why the strategy always works (the champion invariant, the sortedness criterion); and compare strategies by counting elementary operations.

Each outcome comes with a low-stakes source of assessment evidence, detailed in the teacher guide [27]. The fourteen-icon program on the worksheet is a checkable artefact: is every step taken from the manual? would the hand ever hold two bags? A deliberate re-run on fresh bags or straws (the guide's verification phase) tests this ability, separating "it worked once" from "it works always". Stimulus questions have discriminating answers: a child who can say *why* an eliminated bag can never be the heaviest has grasped the invariant, not just the routine. The guide also lists typical failure patterns as diagnostic signals—declaring the straws sorted after a single pass, or having no stopping rule at all—against which a complete pass with no swaps can emerge as an explicit termination criterion for adjacent-pair pass strategies. Finally, cost yields quantitative evidence: the right measure to record is the number of *comparisons*, not of swaps—an already sorted row costs no swaps, yet convincing oneself that it *is* sorted still requires comparisons in the covered game.

The simulators support exactly these practices: counters and preset initial arrangements (random, nearly sorted, reversed) turn "which strategy is better?" into a small fair-test experiment.

6 Conclusion, Limitations, and Future Work

This paper described two primary-school unplugged activities for introducing algorithms as part of informatics education. The clothes-hanger activity turns a familiar object into a simple balance to make maximum search and the champion invariant concrete; the covered-straw activity uses hidden information and a compare-and-swap helper to make sorting strategies observable. In both cases, the central design move is to restrict the executor so that learners must reason with computationally meaningful primitives.

Taken individually, most of the ingredients are known: sorting and weighing are CS Unplugged classics [7], children's ability to sort under hidden information has been established experimentally [38], and block-based environments are widespread. The novelty lies in the architecture that combines them: the joint application of the five principles of Sect. 3 to the first primary years. Children do not mime an algorithm—they program a small physical machine whose state and instruction set are explicit (P1); the constraints live in the material situation rather than in rules the teacher must police (P2); children invent, test, and refine their own strategies instead of memorising named algorithms (P3); the materials are everyday objects of negligible cost (P4); and the companions interpret exactly the activities' primitives, keeping a single abstraction level across physical and digital work (P5). Finally, the same design has been incorporated into a commercial, nationally distributed textbook while remaining fully inspectable through an openly licensed teacher guide and open-source simulators.

The main limitation of this work is that the activities have not yet been empirically evaluated (except for preliminary and informal feedback from the co-authors, the editors, and a small number of teachers). The present paper therefore makes a design argument, not a learning-outcome claim. It is plausible, based on prior CS Unplugged work and recent developmental evidence on hidden-rank sorting, that young learners can engage productively with these tasks, but classroom evidence is still needed.

Future work will test the activities in primary classrooms and in teacher training events, with attention to both learning and feasibility for teachers. Useful data include learners' comparison sequences, verbal explanations, stopping criteria, and changes between physical and simulator-based trials. The straw activity is particularly well suited to logging strategies. Further work should also refine accessibility, examine how teachers conduct the reflection phase, and design a subsequent transition from high-level simulators to age-appropriate programming languages in which learners can implement the strategies themselves.

Acknowledgments. I thank A. Del Zozzo, A. Montresor, and G. Bissoli, co-authors of the booklet and guide; Mondadori Education for the opportunity and for enabling open dissemination of the materials; S. Martini for supporting my research; and R. Davoli, whose balance-scale adaptation of CS Unplugged inspired this work.

Disclosure of Interests. The author has no competing interests to declare that are relevant to the content of this article.

References

1. Bell, T., Alexander, J., Freeman, I., Grimley, M.: Computer science unplugged: school students doing real computing without computers. N. Z. J. Appl. Comput. Inf. Technol. **13**(1), 20–29 (2009)
2. Bell, T., Lodi, M.: Authors' response: keeping the "computation" in "computational thinking" through unplugged activities. Constructivist Found. **14**(3), 357–359 (2019). https://constructivist.info/14/3/357.bell
3. Bell, T., Lodi, M.: Constructing computational thinking without using computers. Constructivist Found. **14**(3), 342–351 (2019). https://constructivist.info/14/3/342.bell
4. Bell, T., Rosamond, F., Casey, N.: Computer science unplugged and related projects in math and computer science popularization. In: Bodlaender, H.L., Downey, R., Fomin, F.V., Marx, D. (eds.) The Multivariate Algorithmic Revolution and Beyond, Lecture Notes in Computer Science, vol. 7370, pp. 398–456. Springer, Berlin, Heidelberg (2012). https://doi.org/10.1007/978-3-642-30891-8_18
5. Bell, T., Vahrenhold, J.: CS Unplugged: how is it used, and does it work? In: Böckenhauer, H.J., Komm, D., Unger, W. (eds.) Adventures Between Lower Bounds and Higher Altitudes, Lecture Notes in Computer Science, vol. 11011, pp. 497–521. Springer, Cham (2018). https://doi.org/10.1007/978-3-319-98355-4_29
6. Bell, T., Witten, I.H., Fellows, M.: CS Unplugged: an enrichment and extension programme for primary-aged children (2015). https://classic.csunplugged.org/books/. Italian translation edited by Renzo Davoli, Giovanni Michele Bianco, and Piergiovanna Grossi
7. Bell, T., Witten, I.H., Fellows, M.: Lightest and heaviest: sorting algorithms. Activity 7 in CS Unplugged (2015). https://classic.csunplugged.org/activities/sorting-algorithms/
8. Bellettini, C., Lonati, V., Monga, M., Morpurgo, A.: To be or not to be... an algorithm: the notion according to students and teachers. In: Proceedings of the 55th ACM Technical Symposium on Computer Science Education V. 1, pp. 102–108. ACM, New York (2024). https://doi.org/10.1145/3626252.3630950 SIGCSE 2024
9. Bissoli, G., Del Zozzo, A., Lodi, M., Montresor, A.: Sulle orme di Milù – Guida di Informatica. Mondadori Education, Milano (2026). https://www.mondadorieducation.it/catalogo/sulle-orme-di-milu-0074279/. teacher guide. ISBN 9791220419833. In Italian

10. Bissoli, G., Del Zozzo, A., Lodi, M., Montresor, A.: Sulle orme di Milù – Informatica. Mondadori Education, Milano (2026). https://www.mondadorieducation.it/catalogo/sulle-orme-di-milu-0074279/. student booklet. In Italian
11. du Boulay, B.: Some difficulties of learning to program. J. Educ. Comput. Res. **2**(1), 57–73 (1986). https://doi.org/10.2190/3LFX-9RRF-67T8-UVK9
12. Brousseau, G.: Theory of Didactical Situations in Mathematics: Didactique des Mathématiques, 1970–1990. Kluwer Academic Publishers, Dordrecht (1997)
13. Capecchi, S., Lodi, M., Lonati, V., Morpurgo, A.: L'informatica negli attuali libri di testo della scuola secondaria di primo grado, alla luce delle Indicazioni Nazionali 2025. In: Bolondi, G., D'Amore, B., Ferretti, F. (eds.) La Didattica della Matematica con e per la scuola, pp. 157–158. Bonomo Editore, San Lazzaro di Savena (2025). incontri con la Matematica no. 39, 2025. In Italian
14. Computer Science Education Research Group, University of Canterbury: Principles – CS Unplugged. https://www.csunplugged.org/en/principles/
15. Crippa, M., Branchetti, L., Lamanna, L., Lonati, V.: Conoscenza procedurale, algoritmi e problemi: spunti di riflessione da una prospettiva interdisciplinare tra matematica e informatica. In: Bolondi, G., D'Amore, B., Ferretti, F. (eds.) La Didattica della Matematica con e per la scuola, pp. 111–112. Bonomo Editore, San Lazzaro di Savena (2025), incontri con la Matematica no. 39, 2025. In Italian
16. Crippa, M., Lonati, V., Lamanna, L., Branchetti, L.: Educare al pensiero algoritmico: riflessioni e proposte per il primo ciclo. In: Pellegrino, M.A., Rossano, V. (eds.) ITADINFO 2025. Terzo Convegno Italiano sulla Didattica dell'Informatica, pp. 19–28. Ledizioni, Milano (2025). in Italian
17. Danielsiek, H., Paul, W., Vahrenhold, J.: Detecting and understanding students' misconceptions related to algorithms and data structures. In: Proceedings of the 43rd ACM Technical Symposium on Computer Science Education, SIGCSE'12, pp. 21–26. ACM, New York (2012). https://doi.org/10.1145/2157136.2157148
18. Feaster, Y., Segars, L., Wahba, S.K., Hallstrom, J.O.: Teaching CS Unplugged in the high school (with limited success). In: Proceedings of the 16th Annual Joint Conference on Innovation and Technology in Computer Science Education, ITiCSE '11, pp. 248–252. ACM, New York (2011). https://doi.org/10.1145/1999747.1999817
19. Futschek, G.: Algorithmic thinking: The key for understanding computer science. In: Mittermeir, R.T. (ed.) Informatics Education – The Bridge between Using and Understanding Computers (ISSEP 2006), Lecture Notes in Computer Science, vol. 4226, pp. 159–168. Springer, Berlin, Heidelberg (2006). https://doi.org/10.1007/11915355_15
20. Gal-Ezer, J., Zur, E.: The efficiency of algorithms: misconceptions. Comput. Educ. **42**(3), 215–226 (2004). https://doi.org/10.1016/j.compedu.2003.07.004
21. Guzdial, M.: Learner-Centered Design of Computing Education: Research on Computing for Everyone. Synthesis Lectures on Human-Centered Informatics. Morgan & Claypool (2015). https://doi.org/10.2200/S00684ED1V01Y201511HCI033
22. Hermans, F., Aivaloglou, E.: To scratch or not to scratch? A controlled experiment comparing plugged first and unplugged first programming lessons. In: Proceedings of the 12th Workshop on Primary and Secondary Computing Education, WiPSCE '17, pp. 49–56. ACM, New York (2017). https://doi.org/10.1145/3137065.3137072
23. Kirsh, D., Maglio, P.: On distinguishing epistemic from pragmatic action. Cogn. Sci. **18**(4), 513–549 (1994). https://doi.org/10.1207/s15516709cog1804_1
24. Lodi, M.: Introducing Computational Thinking in K-12 Education: Historical, Epistemological, Pedagogical, Cognitive, and Affective Aspects. Ph.D. thesis, Alma

Mater Studiorum - Università di Bologna (2020). https://doi.org/10.6092/unibo/amsdottorato/9188
25. Lodi, M., Davoli, R., Montanari, R., Martini, S.: Informatica senza e con computer nella Scuola Primaria. In: Nardelli, E. (ed.) Coding e oltre: l'informatica nella scuola. Lisciani Scuola (2020), in Italian. https://inria.hal.science/hal-02379212
26. Lodi, M., Del Zozzo, A., Montresor, A., Bissoli, G.: Aprite il sussidiario: dalla produzione editoriale a una guida "open" per insegnare Informatica nel primo ciclo. In: ITADINFO 2026. Quarto Convegno Italiano sulla Didattica dell'Informatica (2026). to appear. In Italian
27. Lodi, M., Del Zozzo, A., Montresor, A., Bissoli, G.: Insegnare Informatica – Una guida per docenti del primo ciclo (2026). published in CC BY-NC-SA at https://insegnareinformatica.github.io/. In Italian
28. Lodi, M., Lonati, V.: A few sentences, a whole discipline. Informatics in the new Italian national curriculum guidelines for the first cycle of education. Mondo Digitale (2026). to appear
29. Lodi, M., Martini, S.: Coding, pensiero computazionale e informatica: quale evoluzione possibile? In: ITADINFO 2024. Secondo Convegno Italiano sulla Didattica dell'Informatica, pp. 20–24. Genova University Press, Genova (2024). https://cris.unibo.it/handle/11585/994861, in Italian
30. Lonati, V., Mirolo, C., Monga, M.: Di cosa parliamo quando parliamo di programmi? Mondo Digitale **20**(97), 1–12 (2022). in Italian
31. Mayer, R.E.: Should there be a three-strikes rule against pure discovery learning? Am. Psychol. **59**(1), 14–19 (2004). https://doi.org/10.1037/0003-066X.59.1.14
32. MIM – Ministero dell'Istruzione e del Merito: Indicazioni nazionali per il curricolo della scuola dell'infanzia e del primo ciclo d'istruzione (2025), ministerial Decree of 9 December 2025, no. 221, published in the Gazzetta Ufficiale, General Series no. 21, 27 January 2026. https://www.mim.gov.it/documents/20182/10554370/curricolo_web.pdf. In Italian
33. Monga, M., Davoli, R., Lamanna, L., Forlizzi, L.: Il concetto di algoritmo nei libri di testo: spunti per una riflessione. In: Bolondi, G., D'Amore, B., Ferretti, F. (eds.) La Didattica della Matematica con e per la scuola, pp. 177–178. Bonomo Editore, San Lazzaro di Savena (2025). incontri con la Matematica no. 39, 2025. In Italian
34. Nishida, T., Kanemune, S., Idosaka, Y., Namiki, M., Bell, T., Kuno, Y.: A CS unplugged design pattern. In: Proceedings of the 40th ACM Technical Symposium on Computer Science Education, SIGCSE'09, pp. 231–235. ACM, New York (2009). https://doi.org/10.1145/1508865.1508951
35. Resnick, M., et al.: Scratch: programming for all. Commun. ACM **52**(11), 60–67 (2009). https://doi.org/10.1145/1592761.1592779
36. Sorva, J.: Notional machines and introductory programming education. ACM Trans. Comput. Educ. **13**(2), 1–31 (2013). https://doi.org/10.1145/2483710.2483713
37. Taub, R., Armoni, M., Ben-Ari, M.: CS unplugged and middle-school students' views, attitudes, and intentions regarding CS. ACM Trans. Comput. Educ. **12**(2), 1–29 (2012). https://doi.org/10.1145/2160547.2160551
38. Yang, H.A., Thompson, B.D., Kidd, C.: Children spontaneously discover efficient solutions to a difficult sorting task. Nat. Hum. Behav. **10**, 80–91 (2026). https://doi.org/10.1038/s41562-025-02302-6

Creativity for Computational Thinking

Songs and Other Pedagogical Modes to Teach Computational Thinking in Primary School: An Experience Report

Smitha Balakrishnan[1], Deepali Navghane[2], and Vipul Shah[2](✉)

[1] Tata Consultancy Services, Trivandrum, India
[2] Tata Consultancy Services, Pune, India
{smitha.balakrishnan,deepali.navghane,v.shah}@tcs.com

Abstract. Unplugged approaches to Computational Thinking (CT) are increasingly promoted as equitable and effective alternatives to device-dependent instruction. While prior research demonstrates positive learning outcomes from unplugged CT activities, there is limited reporting on how CT is pedagogically enacted in real classrooms, particularly in low-resource, large-scale contexts. This experience report draws on a pilot carried out with primary students (grades 1–4) in government schools. We describe how CT competencies were taught (through curricular themes and practices) using multiple pedagogical modes, including an inside-out approach that begins with children's daily lived experiences, moves into classroom and subject-based activities, and introduces puzzles and structured challenges only later, before extending to real-life applications. Central to this work is the use of songs and embodied routines to support attention, retention, and participation. We reflect on how these modes fostered participation, habit formation, and transference of CT learning beyond isolated tasks. The experience suggests that creative, low-cost pedagogical modes such as songs and routine activities can play a transformative role in helping young learners see CT as a way of thinking applicable across contexts.

Keywords: Computational thinking · unplugged activities · pedagogical modes · songs · transference of learning · inside-out approach · habit formation

1 Introduction

Computational Thinking (CT) has gained worldwide recognition as a foundational capability for all learners [5,13], and [17]. It is recognized as a way of thinking and problem-solving, a skill as fundamental as the 3Rs - reading, writing and arithmetic. Yet CT is often associated with programming or digital tools, creating barriers in contexts where infrastructure is limited. In such contexts, unplugged approaches offer a practical way to introduce CT without relying on digital devices. In India, the National Education Policy (NEP 2020) [8] explicitly

M. Löffler et al. (Eds.): CMSC 2026, LNCS 16951, pp. 59–72, 2027.
https://doi.org/10.1007/978-3-032-38887-2_5

calls for introducing CT from the foundational years. The National Curriculum Framework (NCF 2023) [11] has integrated CT into mathematics, with emphasis on reasoning, systematic thinking, and problem-solving. However, India is a low-resource, large-scale environment and requires low-cost yet scalable methods to effectively enable CT education as recommended by the national policy.

Systematic assessment of life skills within Indian school systems remains limited, particularly at scale. A growing body of evidence indicates that Indian learners exhibit persistent weaknesses in higher-order cognitive skills, particularly problem-solving and critical thinking. Results from the Program for International Student Assessment (PISA) [12] showed that Indian students performed substantially below the OECD average on tasks requiring non-routine problem-solving, information interpretation and multi-step reasoning. ASER [16] reports consistently note that procedural knowledge does not translate into problem-solving ability, highlighting limitations in reasoning and application. System-level diagnostics reinforce these findings. The World Bank's Strengthening Teaching–Learning and Results for States (STARS) program [21] documents that classroom practices in many Indian states remain dominated by rote-oriented pedagogy, with limited emphasis on analytical reasoning, problem-solving, and higher-order thinking skills.

We address this policy-practice gap through a sustainable unplugged mathematics - integrated CT curriculum aligned to NCF 2023 [11] and ACM India's CSpathshala curricula themes [17]. A key challenge we try to address through this curriculum is how abstract thinking processes such, as CT, can be taught in ways that are intuitive, engaging and transferable for children. We therefore focused on designing suitable pedagogical approaches to create meaningful shifts in how students think. These approaches were embedded into the curriculum through specific pedagogical modes, particularly songs, structured questioning, and an inside-out sequence.

Our aim was to support observable changes in how students think and act. We sought to help students become more systematic in their approach to tasks, break down problems, follow stepwise processes, evaluate their own work and explain their reasoning. We aimed to make students more aware of why they were doing a task, whether their approach was effective, and how it could be improved.

One of the pedagogical modes we used as a means to introduce CT competencies in a way that would be engaging is through songs. Songs and rhymes have long been used as pedagogical tools in early childhood education. Findings [10] demonstrate that music-based pedagogies positively improve cognitive skills. Studies [1] have also shown that music-integrated instruction enhances student engagement and participation in mathematics classrooms. In addition, studies [1,20] and [9] suggest that songs can serve as mnemonic tools that facilitate retention of mathematical concepts, making them a valuable pedagogical tool in early education.

While previous research has examined the role of music in mathematics and language learning, there is little research on how songs can support the teaching of computational thinking. As we developed content on computational thinking, reducing cognitive overload while making learning fun was a key driver. Songs as a pedagogical mode unfolded naturally as a means to achieve this.

Our pilot study in primary schools supported our belief that music or songs can also effectively develop computational thinking skills. This paper offers a reflective account of our experience, highlighting what we observed, what shifted in classrooms, and what we learned about teaching CT ideas effectively to primary-age children. Although this paper is not positioned as a theoretical study, we realized in retrospect that the pedagogical modes described are informed by established perspectives from constructivist learning theory [14] [2], habit formation [3], and the theory of multiple intelligences [4].

This work is presented as an experience report focusing on implementation and classroom processes. Accordingly, our analysis is primarily descriptive, aimed at documenting patterns of engagement and participation, as well as observable shifts in student behaviour. We do not attempt to estimate effect sizes or conduct inferential statistical analyses; instead, the emphasis is on capturing context-rich insights that can inform practice and future studies.

2 The Pilot Experience

2.1 Orientation

Towards achieving the vision of the NEP 2020, the state of Goa, India, expressed its interest in introducing CT education for classes 1 to 4. A one-day orientation workshop for 25 state education department representatives was organized to advocate CT. The outcome of the workshop was the decision to conduct a pilot for classes 1 to 4.

2.2 Pilot Setup

The pilot program was implemented in the three districts of Goa, across 35 government primary schools selected from 12 different blocks (district subdivisions). These schools were selected by the government and involved 1,176 students in grades 1- 4. The pilot was implemented as a fully unplugged program. The participating schools were located in rural villages where resources were limited. Students largely came from lower-middle income and underprivileged communities, including families of migrant labourers. In one instance, a school operated from a rented building originally intended as a storage facility, as the main school infrastructure was in a dilapidated condition. In several schools, classrooms were shared spaces across grades or repurposed as staff rooms.

2.3 Teacher Training

The selected teachers from the participating schools underwent an intensive two-day training program. The program, conducted largely in Marathi, the local language, began with an introduction to computational thinking, its relevance in the current times, and its applicability in their own lives. Initial discussions revealed that many teachers associated CT primarily with computers. Therefore, we included a segment that focused on clarifying the broader conceptual foundations of CT and its relevance to everyday problem-solving. This process helped shift teacher perceptions and created openness to subsequent learning.

Songs were an integral part of the training program. In fact, as part of the workshop activities teachers designed contextualized lesson plans where they created their own songs and incorporated them into their presentations. The songs were simple, repetitive, and often set to familiar tunes, which facilitated ease of learning and recall.

We provided sustained teacher support throughout the five-month pilot through a series of structured mentoring interactions. Teachers participated in biweekly post-workshop connect sessions that we called the Manthan Series. These sessions focused on reflecting on classroom experiences, addressing challenges, responding to teacher queries, and supporting preparation for upcoming sessions. In addition, ongoing informal support was facilitated through a dedicated WhatsApp group that helped teachers share experiences, seek timely assistance, and engage with peers and our team as needed.

The pilot focused on three computational thinking themes from [11] and [17]: Systematic Listing, Counting and Reasoning, Modelling, and Following and Devising instructions (Algorithms - stepwise thinking and procedures).

2.4 Pedagogical Modes Used

While prior work [17] had already established the key themes for CT in school education, our focus was on how these ideas are experienced, how children engage with them, revisit them, and internalize them through pedagogy, treating pedagogy as the primary medium of communication.

A defining feature of our unplugged approach was the deliberate use of multiple pedagogical modes to make CT accessible and memorable for young learners. To create a holistic impact in the way students perceived, thought and processed everything around them, our pedagogical modes were grounded in the following philosophies:

- Songs as Communication Tools and Embodied Routines
- An Inside Out Sequence
- Questions as Meaning-Making Anchors

2.4.1 Songs as Communication Tools and Embodied Routines Studies show that music engages not only perception but also higher-order cognitive functions like attention, memory, and emotional regulation [9,15] and [18]. In classroom contexts, the use of songs has been shown to significantly increase primary students' motivation, attention, and willingness to participate in learning activities [18], complementing evidence that musical structures support memory and sequencing [9].

Songs were therefore used throughout the intervention to communicate and reinforce CT ideas. They supported memory, sequencing and recall of CT themes such as systematic listing, and stepwise reasoning as suggested by [1] and [20]. Short, repetitive songs with simple language were sung regularly before, during or at the end of activities. Teachers also used lines from the songs to reinforce learning during classroom interactions.

The primary school teachers participating in the pilot were already familiar with and comfortable using songs, nursery rhymes, rhythmic chanting, and poetry recitation in their teaching practices. Such approaches are deeply embedded in elementary education in India, particularly in rural and semi-rural contexts. NEP 2020 and NCF 2023 [8] promote the use of songs, rhymes, and music in early-grade pedagogy as part of a play-based, arts-integrated approach. As a result, teachers readily adopted this approach in the classroom and were pleasantly surprised to discover that CT concepts could also be effectively taught through this medium.

They introduced songs and chants such as *"Every time you do something, do it systematically"* into routine instructions to reinforce habits of stepwise thinking. Songs were used to teach classification, finding words in a word grid, and pattern recognition (e.g., *"Sparrow bhurr, crow bhurr"* games for grouping and sorting). Some teachers also improvised on the tunes while others translated them into other languages, such as Hindi, to meet the language needs of their students. These songs functioned as cognitive and emotional anchors, helping retention and internalization of the CT ideas without requiring any other formal definition.

Table 1 presents two songs used during the pilot, each translated into and taught in the local languages Marathi and Hindi. The song *"Every time we do something, do it systematically"* was introduced first and became particularly popular among students. Teachers reported that words such as "systematically" and its local equivalent *"krumma krummane"* (step-by-step) were readily adopted into students' everyday vocabulary, suggesting that they may have functioned as anchors that supported internalization of computational thinking habits. The second song, *"Look Across Each Row,"* was used to help students locate letters and words within word-search grids. Accompanied by simple finger actions to indicate rows, columns, and crosswise movement, the song helped children understand how to search systematically without missing any letters or words. This practice later helped with similar grid-based tasks, such as identifying odd letters, filling in number grids, and solving puzzles like Sudoku and KenKen.

Teachers reported that students found the songs rhythmic and enjoyable, and when accompanied by simple actions, they reduced hesitation and anxiety around 'thinking tasks,' fostering motivation and voluntary participation. According to teachers, this enabled young learners to engage more deeply with computational thinking as a way of thinking, rather than merely a set of abstract skills.

What we did not anticipate was that the songs became behavioural cues and not just content carriers. For many children, the song itself became a sign to

Table 1. CT songs used in the pilot

Systematically Song	Words Search Song
Every time you do something,	Look across each row
Do it systematically	Look across each row
Every time you do something,	Left to right and right to left,
Do it systematically	Look across each row
	Are we good to go?! **No**...
Step by Step, Step by Step	
That's doing it systematically	Look through every column
Step by Step, Step by Step	Look through every column
That's doing it systematically	Up and down; top to bottom,
	Look through every column
When your task is big and tall,	Are we good to go?! **No**...
Don't you worry, don't you stall	
Break it into smaller parts	There's one more way to go!
That's the easy way to start	There's one more way to go!
	Crosswise, Slantwise, Corner to Corner
Every time you do something,	That's one more way to go!
Do it systematically	Are we good to go?! **Yes**...
Every time you do something,	
Do it systematically	

slow down, organise their actions, and reflect on steps. For instance, teachers observed that students not only remembered the 'do it systematically' song but also applied it across activities, often pausing mid-task to note the need to work systematically and think through their approach. They began to 'notice' unorganised instances and voluntarily reported the need for being 'systematic'. In another example, a teacher reported that students in her class observed that their classroom cupboard was unorganised and suggested organising it themselves.

2.4.2 An Inside-Out Sequence The inside-out sequence was a central component of our unplugged design, grounding CT learning in everyday experiences. Children aged 6 to 11 are naturally curious, eager to learn, explore, and experiment [6,7]. At this stage, they have fewer assumptions and biases, as their independent thinking is still developing and their ideas are not yet firmly formed [14]. We saw this as an opportunity to leverage their natural abilities and encourage their 'Thinking' skills. Our aim was to help them become aware of not just what they think, but also of what they are capable of thinking and noticing. We encouraged them to observe and reflect on the things they do every day.

The effectiveness of an **inside-out approach** to sequencing activities that introduced computational thinking was a key insight from this experience. Rather than beginning with puzzles or formal problems, learning started from what children already do in their daily lives.

The sequence of activities unfolded as follows:

- **Daily life activities**: Any CT idea was first introduced through activities that prompted children to reflect on familiar routines, such as brushing their teeth, packing a bag, or arranging footwear. Reflection was a key stage in every activity discussion. During this stage, the teacher, in the role of a facilitator, used guiding questions to prompt students to reflect on the processes involved in these everyday activities. These prompts helped students think about everyday actions in a structured way and explain their reasoning. Most of the reflection took place through classroom discussions, and in some instances, teachers also encouraged students to write down their thoughts. Teachers reported that eventually students began to share their experiences more actively, recall and identify steps they had missed, and relate their learning to new situations at home or elsewhere. For instance, a little boy spoke proudly as he shared his experience of having organized a cupboard that his mother, who worked as domestic help, had brought home. He was able to fit his and his siblings' clothes into the cupboard because he organized them systematically.
- **Classroom and subject-based activities**: The same thinking was later applied to mathematical tasks and classroom organization. For instance, in mathematical tasks, children followed a systematic, step-by-step method to add numbers, find missing numbers, or think of all possible ways of getting a number as a sum. They voluntarily used the same methods to organise their classroom - the teacher's table, books on a table, the classroom cupboard or reading corner, or even queuing up for their mid-day meal.
- **Puzzles and structured challenges**: Only after habits of systematic thinking were established, were puzzles introduced. Children applied the same methods to solving crosswords, number grids, counting shapes within a shape or magic triangles.
- **Real-life scenario applications**: At the end of each lesson, children were given take-home assignments where they were encouraged to notice and apply CT in less structured, real-world situations such as observing how their wardrobes were organised and why they were organised so, how their kitchens were organised and why so. This extended to short field visits to a stationery shop or a grocery store as well.

We observed that when CT is introduced first through puzzles, children often treat it as a special kind of activity, relevant only within the puzzle context. In contrast, starting from immediate, relatable experiences such as their daily routines and gradually moving to classroom experiences and puzzles helped children recognize CT as something they already use and can consciously refine.

In retrospect, we realised that our approach aligns with the constructivist learning theory [14] and [2] that emphasises that children create understanding most effectively when new ideas are grounded in familiar contexts drawn from their everyday lives. The gradual sequencing of activities aided progressive abstraction supporting stronger transfer of learning beyond specific tasks for young learners.

2.4.3 Questions as Meaning-Making Anchors A key shift emphasized during the pilot involved teachers moving from explaining solutions to facilitating thinking through questions, adopting a more inquiry-based approach. Teachers consistently prompted reflection using questions such as: *What were you trying to do? How did you do it? What are the steps involved? What did you do first? What comes next? Have you considered all possibilities? Did you miss anything? Could this be done differently? How do you know you have completed the task?* These questions functioned as cognitive anchors, guiding students' thinking, gently redirecting it when necessary, and fostering discovery. They also sustained students' attention and promoted inquiry and curiosity. Repeated across lessons, questioning became a shared classroom practice. Teachers were encouraged to reframe any impulse to explain or instruct into a question, marking a deliberate shift from telling to asking.

We observed that this shift altered classroom dynamics: children spoke more frequently, justified their ideas, and assumed greater ownership of their learning. One teacher illustrated this change through an example involving her students preparing for a scholarship exam. When the teacher was once unexpectedly absent, and the children encountered a problem they could not solve, they did not wait for the next class, as was their usual practice. Instead, they proactively sought out another available teacher and asked for guidance, demonstrating increased ownership, confidence, and problem-solving initiative.

Teachers also reported that children began to develop a greater interest in subjects they once were apprehensive about, such as Mathematics or English. They observed that question prompts shifted the focus from arriving at a fixed solution to exploring methods and reasoning, while also making the classroom a safer space for children to think aloud and ask questions.

2.5 Observed Transformation

A recurring theme in this experience was the difference between short-term change and long-term transformation. Rather than isolated improvements in task performance, the teachers observed gradual shifts in how children approached activities. Over the period of five months, data from teacher surveys, the Manthan sessions, and the WhatsApp group interactions indicated that teachers noticed children becoming more organized, more willing to explain their thinking, and more confident in trying different approaches.

Importantly, these behaviours were reported not only during CT-focused lessons but also across other subjects and everyday classroom routines. We infer that the songs and repeated everyday activities likely contributed to this shift by reinforcing habits consistently and joyfully, enabling CT processes to gradually shift from conscious effort to more automatic responses.

2.6 Observed Outcomes

For our pilot, data was collected through teacher surveys and a pre- and post-assessment administered for students. Prior consent was obtained from participating teachers before data collection. Teacher surveys were conducted using Google Forms, which were shared through WhatsApp groups. The following results illustrate the general trends observed throughout the pilot phase.

2.6.1 Assessment The pre- and post-assessments were conducted offline in classrooms, in the local language (Marathi). Teachers projected the questions for students. Designed for a regular 45-minute class period, we also had to account for teachers needing to read aloud or explain questions. For the same reason, the assessments were limited to six questions for Grades 1–2 and nine for Grades 3–4.

Modelled on Bebras tasks [19], the assessments included easy, medium, and difficult multiple-choice questions with one correct answer. The assessments evaluated both students' understanding and CT skills. The questions specifically assessed key competencies such as decomposition, systematic information processing, pattern recognition, abstraction, and algorithmic thinking, which were introduced through the curricular themes including, systematic listing, modelling, and following and devising instructions, making CT concepts more meaningful for both teachers and the education department.

Teachers were given a template to record scores through the WhatsApp group. Some used the Excel format, while others recreated it on paper and recorded scores manually. Photos of these records were shared with us and compiled into a consolidated Excel dataset for analysis.

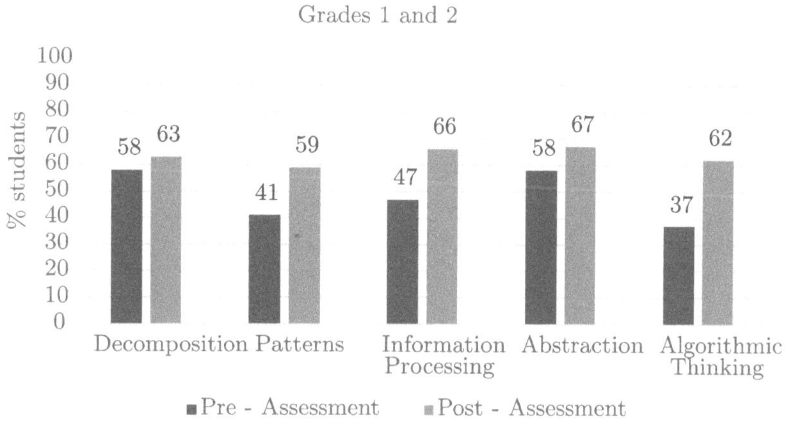

Fig. 1. Grades 1 & 2.

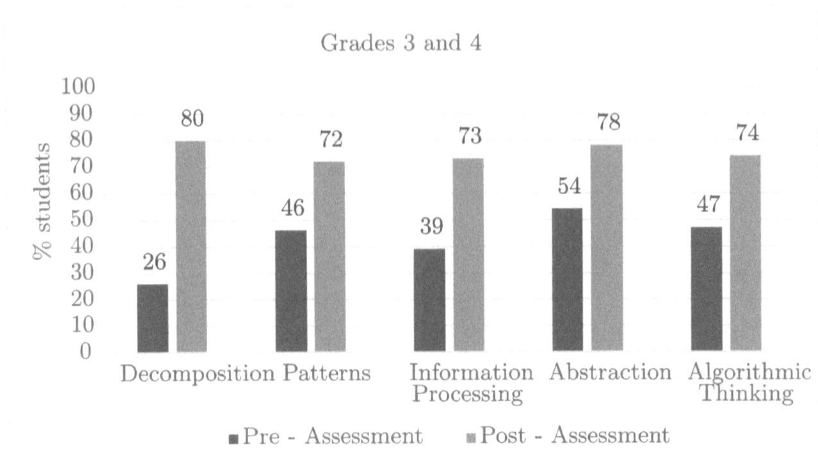

Fig. 2. Grades 3 & 4.

While the pilot covered a total of 1,176 students, the number of students who participated in the pre- and post-assessments varied. In Grades 1–2, 132 students completed the pre-assessment and 216 the post-assessment; in Grades 3–4, 227 students completed the pre-assessment and 216 the post-assessment. For our analysis, we considered only the subset of students who completed both assessments, comprising 127 students in Grades 1–2 and 114 in Grades 3–4.

43% students in Grades 1–2 were able to answer the pre-assessment questions correctly. The number of students who could answer the post-assessment questions correctly was at 58%. For grades 3–4, 39% students answered the pre-assessment questions correctly. The number jumped to 71% for the post-assessment. Figures 1 and 2 show student performance in grades 1–2 and grades 3–4, respectively.

These results are presented as descriptive indicators of change within the pilot cohort. Given the absence of a control group, the short duration of the intervention, and the exploratory nature of the study, we do not interpret these differences as causal effects.

The results indicate early improvements, suggest the potential impact of the approaches used, and provide a basis for future refinement and evaluation. Longitudinal studies would be required to evaluate long-term retention and transfer of CT skills.

2.6.2 Teacher Survey Questions We asked teachers the following questions as part of the survey:

- Which of the following activities did you conduct in your classrooms? (List of classroom activities)
- What aspects of CT session implementation do you find beneficial? (Select all that apply)

- Increased student engagement and keenness to learn.
- Enhanced teamwork and collaboration amongst students.
- Improved critical thinking skills in students. They are more observant, curious and willing to ask questions.
- Improved ability to solve problems using CT concepts.
- Positive behavioural changes in students such as being systematic, organized and confident.

– How did your students find the CT activities?
– What changes do you see in yourself after learning computational thinking in the workshop and teaching it in the class?
– What are your thoughts about CT after implementing this program in your classroom?
– Write about your most meaningful experience where you think CT made a deep impact on students
– Did you observe application of transfer of CT skills to other subjects or other areas of your work?

Student Engagement and Behavioural Change
Teacher survey results indicated consistently high levels of engagement with 96.5% reporting students found CT activities engaging, 92.9% observing increased attentiveness and keenness to learn, 89.3% noting positive behavioural changes such as confidence, organization, and systematic habits. Survey responses also indicated that students became more observant, curious, and willing to ask questions: behaviours teachers associated with CT reasoning.

Transfer of Learning Across Contexts
Beyond these behavioural shifts mentioned in the Sect. 2.6.2, another key observation during the pilot was the transfer of CT practices across contexts. Teachers reported using CT strategies to teach other subjects, including language and mathematics. They also reported using it to organise their own professional and personal work, including handling classroom challenges and everyday problem-solving. Students were also reported to have applied stepwise reasoning at home and in everyday tasks. In one school, teachers reported an increase in enrolment, which they attributed to observable changes in classroom culture and student confidence, communicated informally through parent word-of-mouth.

2.7 Teacher and Parent Testimonials

Teachers themselves described a fundamental shift in their mindsets:

– "It's a joy of being a computational thinking teacher and the satisfaction in finding the master key to learn and teach in a systematic way." – Smita Mahesh Raul, GPS Punarvasan, Goa
– "Students applied the learning in the other tasks at home that they did in the classroom such as keeping things in place, keeping them organized, keeping

themselves and other things in the house in order, teaching the people in the house to keep things in an organized manner. Teachers don't have to give them instructions anymore. Children do it by themselves." Arati Kolambkar, GPS Chaudi Canacona, Goa

- "A parent came and said that their son was not packing his school bag, but has now started packing the bag properly. A 4th grade student was forgetting the sequence while multiplying two-digit numbers. When he was taught the algorithm, the same boy started telling the steps by himself, and other students also began solving the multiplication by telling each other the steps properly." - Shweta Joshi, S.S. Samiti Primary School, Dhavali, Ponda, Goa

3 Limitations

As an experience report, this paper prioritizes reflective insights from classroom practice drawn from government primary schools in Goa, representing low-resource, rural and semi-rural Indian contexts. In this report, we considered qualitative change and transfer. We are also well aware that much of the evidence presented comes from teacher observations, informal feedback, and reflective accounts. While these perspectives are valuable, they may be subject to bias, enthusiasm effects, or selective reporting.

Moreover, our observations are based on the short duration of the pilot and implementation during and immediately after the pilot period. While there are a few anecdotal instances of long-term sustainability of habit and CT skills over the years, generalisability of our observations is limited and requires further study and research. The pedagogical modes described are those suited to early and foundational CT competencies and how these would need to evolve for advanced competencies is not addressed.

4 Conclusion

The pilot demonstrates that computational thinking can be effectively cultivated through the themes and practices of systematic thinking and reasoning, using pedagogical modes such as songs, embodied routines, and other unplugged, hands-on activities in low-resource classrooms. By focusing on pedagogy and reasoning, the initiative achieved high engagement, observable behavioural shifts, and meaningful transfer of learning across subjects. This experience contributes practical insights into how CT can be made visible, teachable, and scalable in low-resource environments at the foundational stages.

References

1. An, S., Capraro, M., Tillman, D.: Elementary teachers integrate music activities into regular mathematics lessons: effects on students' mathematical abilities. J. Learn. Arts: A Res. J. Arts Integrat. Schools Commun. **9** (2013). https://doi.org/10.21977/D99112867

2. Bruner, J.S.: The Process of Education. Harvard University Press, Cambridge (1960)
3. Fiorella, L.: The science of habit and its implications for student learning and well-being. Educ. Psychol. Rev. **32** (2020). https://doi.org/10.1007/s10648-020-09525-1
4. Gardner, H.: Frames of Mind: The Theory of Multiple Intelligences. Basic Books, New York (1983)
5. Grover, A.S.: Computational thinking: the idea that lived. Blog@CACM (2024). https://cacm.acm.org/blogcacm/computational-thinking-the-idea-that-lived/
6. Gruber, M.J., Fandakova, Y.: Curiosity in childhood and adolescence-what can we learn from the brain. Curr. Opin. Behav. Sci. **39**, 178–184 (2021). https://doi.org/10.1016/j.cobeha.2021.03.031. https://www.sciencedirect.com/science/article/pii/S2352154621000796
7. Jirout, J., Evans, N., Son, L.: Curiosity in children across ages and contexts. Nat. Rev. Psychol. **3** (2024). https://doi.org/10.1038/s44159-024-00346-5
8. Kasturirangan, K., et al.: National Education Policy (2020). https://www.education.gov.in/sites/upload_files/mhrd/files/NEP_Final_English_0.pdf
9. Knott, D., Thaut, M.H.: Musical mnemonics enhance verbal memory in typically developing children. Front. Educ. **3** (2018). https://api.semanticscholar.org/CorpusID:46887572
10. Kumara, G., Herath, J.: Integrating music activities into mathematics teaching to enhance cognitive skills in primary education: a systematic literature review. Int. J. Res. Innov. Social Sci. **9**, 9589–9597 (2025). https://doi.org/10.47772/IJRISS.2025.903SEDU0728
11. National Council of Educational Research and Training: National Curriculum Framework for School Education (2023). https://ncert.nic.in/pdf/NCFSE-2023-August_2023.pdf
12. Organisation for Economic Co-operation and Development: Pisa 2009 results: What students know and can do. Technical report, OECD Publishing, Paris (2010). https://www.oecd.org/pisa/pisaproducts/48852548.pdf
13. Palop, B., Díaz, I., Rodríguez-Muñiz, L.J., Santaengracia, J.J.: Redefining computational thinking: a holistic framework and its implications for k-12 education. Educ. Inf. Technol. **30**, 13385–13410 (2025). https://api.semanticscholar.org/CorpusID:275832634
14. Piaget, J.: The origins of intelligence in children. International Universities Press (1952). translated by Margaret Cook
15. Piccardi, L., Palmiero, M., Nori, R.: The role of sounds and music in emotion and cognition. Brain Sci. **14**(3) (2024). https://doi.org/10.3390/brainsci14030192. https://www.mdpi.com/2076-3425/14/3/192
16. PRATHAM: Annual status of education report (rural) 2022 (2023). https://asercentre.org/wp-content/uploads/2022/12/ASER-report_2022-1.pdf
17. Ramanujam, R., Shah, V.: Curricular choices for computational thinking in large scale low resource environments. In: Fernau, H., Schwank, I., Staub, J. (eds.) Creative Mathematical Sciences Communication, pp. 117–129. Springer, Cham (2025). https://doi.org/10.1007/978-3-031-73257-7_9
18. Şener, S., Erkan, D.: The effect of songs on primary school students' motivation. Int. Online J. Educ. Teach. **5**(4), 867–875 (2018). http://iojet.org/index.php/IOJET/article/view/454
19. Shah V, G.S.: Bebras India (2023). https://www.bebras.in

20. Trinick, R., Ledger, G., Major, K., Perger, P.: More than counting beats: connecting music and mathematics in the primary classroom. Int. J. Math. Teach. Learn. **17** (2016). https://doi.org/10.4256/ijmtl.v17i3.32
21. World Bank: India strengthening teaching learning and results for states (stars): program appraisal document. Technical report, World Bank, Washington, DC (2020). https://documents.worldbank.org/

CodyRoby: An Embodied Programming Environment for Computational Thinking

Alessandro Bogliolo[1,2](✉)

[1] DiSPeA and AIMRC, University of Urbino, 61029 Urbino, Italy
[2] DIGIT srl, 61029 Urbino, Italy
alessandro.bogliolo@uniurb.it

Abstract. Unplugged activities are widely used to introduce computational thinking without digital technologies, but they are often regarded as educational tools rather than programming environments. This paper presents CodyRoby as an embodied programming environment that combines a programming language, an execution model, and a physical environment in which programs can be executed, observed, tested, and discussed. The CodyRoby language includes elementary movement instructions, selection constructs, and iteration constructs supporting fixed-count, goal-oriented, and condition-controlled repetition. Programs are represented as sequences of cards and executed by a human robot acting on instructions provided by a human programmer. Through a small set of visual conventions based on colors, symbols, and indentation, CodyRoby supports sequence, selection, iteration, and nesting. We argue that CodyRoby externalizes program execution through embodied interaction. Program state, control flow, sensing activities, debugging, and execution verification become directly observable through tangible artifacts and collective participation, providing a simple and transparent framework for teaching, communicating, and exploring computational concepts. Building on more than a decade of educational practice and dissemination involving tens of thousands of teachers, the paper provides a systematic description of the CodyRoby language and execution model and discusses their relevance for embodied programming environments and computational thinking.

Keywords: Embodied Programming · Computational Thinking · CodyRoby

1 Introduction

Programming languages and programming environments are designed to enable the specification and execution of computational processes. While they make it possible to create increasingly complex programs, many of the mechanisms involved in program execution remain hidden from learners. State transitions, condition evaluation, control flow, iteration, and decision making are typically

M. Löffler et al. (Eds.): CMSC 2026, LNCS 16951, pp. 73–86, 2027.
https://doi.org/10.1007/978-3-032-38887-2_6

performed internally by the machine and become visible only through their effects.

The growing importance of computational thinking as a fundamental competence has stimulated the development of educational approaches aimed at making computational concepts accessible to diverse audiences [19]. Among these, unplugged approaches have played a particularly important role by removing technological barriers and introducing computational concepts through physical activities, games, and manipulatives. Since the pioneering work of Bell and colleagues on CS Unplugged [1,3], numerous activities have demonstrated that fundamental concepts of computer science can be effectively explored without computers [2]. Studies on embodied cognition and embodied learning have also highlighted the importance of physical interaction and bodily experience in the construction and communication of abstract concepts [9,10,18].

The literature recognizes board and card games, together with embodied navigation activities on discrete grids, as important approaches for fostering computational thinking [7]. However, these unplugged activities are presented primarily as educational tools for introducing specific concepts rather than as programming environments in their own right.

This paper presents CodyRoby as an embodied programming environment for computational thinking. Originally conceived as an unplugged role-playing activity, CodyRoby combines a programming language, an execution model, and a physical environment in which programs can be executed, observed, tested, and debugged. The language includes primitive movement instructions together with control structures for selection and iteration, while the execution model explicitly separates the role of the programmer (Cody) from that of the executor (Roby).

The central idea behind CodyRoby is that computational processes should not only be represented but also enacted. Program execution becomes a visible and collective activity in which state, control flow, condition evaluation, and execution traces are externalized through physical artifacts and embodied actions. As a result, concepts that are normally hidden inside a computer or a robot become directly observable and discussable.

The contribution of this paper is primarily descriptive. Rather than evaluating learning outcomes, we provide a systematic presentation of the CodyRoby language and environment, describing its syntax, semantics, and execution model. We show how sequence, selection, iteration, and nesting are represented through a small set of coherent visual constructs and how their execution is embodied through the interaction between Cody, Roby, and the environment.

The work builds on more than a decade of educational practice and dissemination involving over 40,000 teachers and numerous classroom, outreach, and public-engagement activities. By documenting the design principles and computational structures underlying CodyRoby, the paper aims to establish a foundation for future research on embodied programming, computational thinking, and the communication of computational ideas.

1.1 Previous Work

Helping novice programmers understand how programs are executed has long been recognized as a fundamental challenge in computer science education. Du Boulay, O'Shea, and Monk introduced the concept of the *notional machine*, arguing that learners require a simplified yet faithful mental model of program execution in order to reason about computational processes. Their "glass box" metaphor emphasizes that educational programming environments should make the mechanisms underlying program execution transparent while hiding unnecessary implementation details [6].

Several educational programming environments have addressed this challenge by simplifying different aspects of programming. Karel the Robot [14] introduced programming through the navigation of a robot in a discrete grid, demonstrating that algorithmic thinking can be developed within a simple execution domain while preserving the essential constructs of imperative programming. Logo and Turtle Geometry [12,13] complemented this approach by exploiting body-syntonic reasoning, encouraging learners to understand computation by identifying with the movements of the turtle and relating abstract instructions to their own bodily experience.

Another important line of work has focused on simplifying the syntax of programming languages. Block-based programming environments [17], such as *Scratch* [15] and *Blockly* [8], replace textual syntax with visual blocks that make program structure explicit while reducing syntactic barriers. Although these environments greatly simplify program construction and highlight instructions during execution, program execution still takes place inside a computer.

Within the unplugged tradition [7], activities such as *KidBots*, developed as part of CS Unplugged [3], demonstrate that programming concepts can be introduced through embodied role playing, where participants act as programmers and robots. These activities make programming a collaborative physical experience and eliminate the need for digital devices, but they are primarily presented as educational activities rather than as programming environments with explicitly defined languages and execution models.

CodyRoby builds upon these complementary traditions. Like Karel, it adopts robot navigation on a discrete grid as its computational domain; like Logo, it exploits body-syntonic reasoning; like block-based languages, it employs a visual syntax; and like KidBots, it externalizes program execution through embodied interaction. Its distinguishing contribution is the systematic integration of these ideas into an embodied programming environment whose language, execution model, and physical artifacts collectively externalize the notional machine, making computational processes directly observable, discussable, and verifiable.

The starter kit of CodyRoby, together with the first activities, was proposed in 2014 as a do-it-yourself instrument to foster the participation of Italian schools in Europe Code Week [4]. Repetition and selection constructs were introduced in 2016 [5].

2 The CodyRoby Programming Environment

CodyRoby comprises both a programming language and a programming environment. The language provides primitive movement instructions together with control structures for sequencing, selection, and iteration. The environment consists of a discrete grid, a robot, obstacles, targets, and sensing mechanisms.

The language is designed to be functionally complete with respect to the environment. All actions required to navigate the environment can be expressed through combinations of the available instructions and control structures. Additional action primitives can be introduced for specific activities, but they are not required for the navigation tasks considered in this work.

2.1 Environment and State

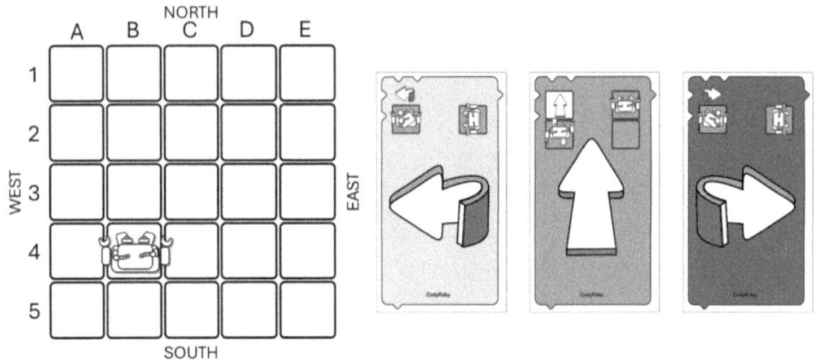

Fig. 1. From left to right: the CodyRoby 5 × 5 grid and the three cards of the elementary instruction set (turn left, move forward, and turn right).

The CodyRoby environment is based on a discrete square grid. While the language itself does not impose constraints on the size of the grid, the standard CodyRoby environment adopts a 5 × 5 grid (Fig. 1). This size was chosen because it represents the smallest grid that allows the construction of non-trivial programming tasks and activities, while remaining easy to reproduce and manipulate in educational settings.

The grid is identified through column labels (A–E) and row labels (1–5). The orientation of Roby is represented through the four cardinal directions North (N), East (E), South (S), and West (W). A state can therefore be identified by the triple (x,y,o), where x denotes the column, y the row, and o the orientation. For example, the state (B,4,N) identifies a robot located in cell B4 and facing North.

While coordinates and orientations can be used to identify states unambiguously, CodyRoby primarily relies on a visual representation. The shape of the robot makes its orientation immediately recognizable, eliminating the need for additional directional markers and supporting a direct interpretation of movement instructions. As a consequence, the state is usually perceived and manipulated visually rather than symbolically.

This representation is preserved across different implementations of the environment, including role-playing activities, board-game versions, and digital simulations. In all cases, position and orientation remain directly observable properties of the execution state.

2.2 Primitive Instructions

The elementary instruction set consists of three textless cards representing the actions Move Forward (green), Turn Left (yellow), and Turn Right (red), as shown in Fig. 1.

The cards use colors, arrows, and shapes rather than written keywords, making the language independent of natural languages and accessible before literacy develops. The semantics of the instructions are relative to the current orientation of Roby. Consequently, orientation forms an essential part of the execution state, as in turtle-graphics systems and mobile robots [12].

The instruction set intentionally focuses on navigation. More complex movements can be expressed through combinations of primitive instructions. For example, moving backward can be obtained through a sequence of rotations and forward movements. This minimal set of primitives is sufficient to support all navigation tasks considered within the CodyRoby environment.

2.3 Program Composition

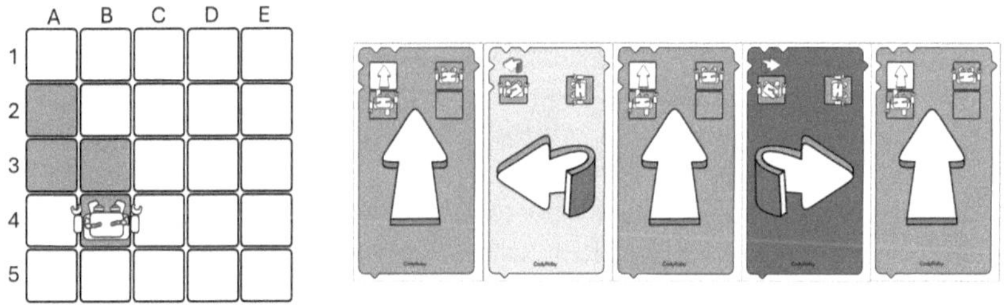

Fig. 2. A linear code segment and the corresponding trajectory, bringing Roby from the initial state (B,4,N) to the final state (A,2,N).

Programs are constructed by arranging instruction cards from left to right. The resulting sequence represents the order in which instructions are executed.

The shape of the cards supports this composition mechanism. Each card has a slightly recessed left edge and a slightly protruding right edge, suggesting how consecutive instructions should be aligned and connected. This physical affordance provides an intuitive introduction to sequential composition and contributes to making the syntax of the language directly manipulable.

A CodyRoby program is therefore represented as an ordered sequence of instructions whose execution follows the same left-to-right order in which the cards are arranged.

2.4 Programs, Executions, and Trajectories

Executing a program transforms the state of Roby and generates a trajectory on the grid. Figure 2 illustrates a simple program together with the corresponding trajectory.

The trajectory provides an intuitive geometric representation of the spatial effects of computation.

A CodyRoby computation can be described through three complementary representations. The first is the program itself, represented as an ordered sequence of instruction cards. The second is the execution, represented by the succession of states assumed by Roby during program execution. The third is the trajectory generated on the grid, i.e. the spatial effect of computation. Highlighting the cells traversed by Roby allows learners to predict, verify, and discuss the outcome of a program.

These representations are related but not equivalent. The program describes the instructions to be executed, the succession of states describes how the execution evolves over time, and the trajectory captures the resulting movement in space. Together, they provide complementary perspectives on the same computational process.

3 Control Structures

The elementary instructions introduced in the previous section allow the construction of linear programs. More expressive programs require mechanisms for repetition and decision making. CodyRoby supports these capabilities through visual constructs for iteration and selection.

A distinctive feature of the language is the uniformity with which control structures are represented. Repetition and selection are both expressed through dedicated blue control cards, while conditions are represented by reusable symbols, and the scope of each construct is identified through indentation. As a result, a small set of graphical conventions is sufficient to represent the fundamental structures of imperative programming.

3.1 Repetition

Iteration is represented through a single Repeat card. The specific semantics of the loop are determined by the symbols placed on the card, allowing the same construct to support different forms of repetition.

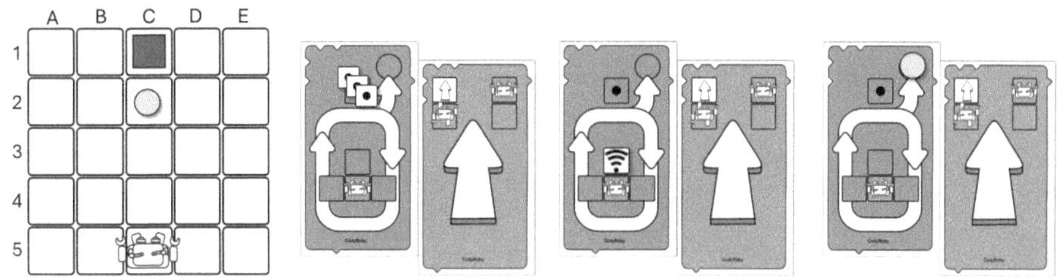

Fig. 3. Given the environment and the initial state (C,5,N), the figure shows three equivalent forms of repetition that can be used to bring Roby to the final state (C,2,N).

Figure 3 illustrates the three main variants of repetition in CodyRoby through a simple navigation task. Starting from the same initial state (C,5,N), Roby must move three cells forward until reaching the target state (C,2,N). The three programs generate the same trajectory but rely on different strategies.

In the first program (*repeat* a fixed number of times) the number of iterations is specified explicitly through three unit markers placed on the Repeat card. In the second (condition-based repetition), iterations continue *while* a free cell is available in front of Roby (the dark square represents a block that prevents Roby from entering the cell). In the third (target-oriented repetition), iterations continue *until* the target (i.e., a yellow circle) is reached.

Although these forms of repetition correspond to different programming constructs, they are all represented by the same Repeat card. The loop symbol on the card denotes the iterative nature of the construct, while three distinct positions are provided for the attached symbols that specify the continuation or termination criterion. Iteration units placed above the loop identify a count-controlled repetition; a sensor symbol placed within the loop identifies a condition-controlled (while) repetition; and a target symbol placed in the upper-right corner identifies a goal-oriented (repeat-until-target) repetition.

3.2 Scope and Indentation

The body of a repetition construct is identified through indentation. Instructions placed at a deeper indentation level belong to the loop body and are repeatedly executed according to the condition specified by the Repeat card.

This mechanism provides a visual representation of scope without requiring delimiters, brackets, or explicit end markers. The same indentation rule applies independently of the type of repetition being used. In Fig. 3, the indented Move Forward instruction identifies the body of the loop in all three programs. The consistency of this representation contributes to the uniformity of the language and makes the relationship between control structures and controlled instructions immediately visible.

3.3 Nested Repetition

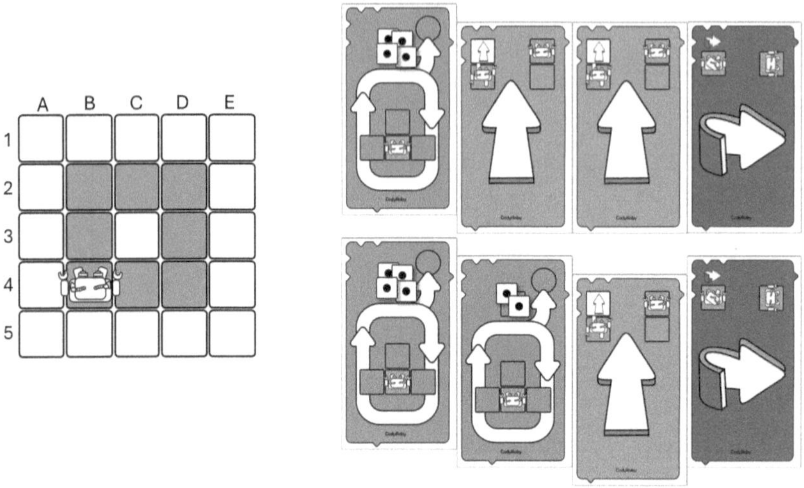

Fig. 4. Two different ways of bringing Roby along a square path, with and without nested loops.

Repetition constructs can be concatenated and nested. Additional indentation levels identify nesting relationships among control structures and allow repeated patterns to be expressed at different levels of abstraction. Figure 4 illustrates two equivalent programs generating the same square trajectory. The nested version makes explicit a repeated pattern within another repeated pattern, exposing regularities that remain implicit in the linear representation.

Through nesting, CodyRoby supports hierarchical program organization while preserving the visual simplicity of its syntax. Although CodyRoby is mainly intended to be used to write short programs with two or three nested constructs, the syntax does not impose any inherent limitation on nesting depth or code length.

3.4 Selection

Conditional execution is represented through blue *if* and *else* cards. As in repetition constructs, conditions are specified through symbols attached to the control card. The same sensor symbols used in condition-controlled loops can therefore be reused in conditional statements, providing a uniform representation of conditions throughout the language.

The if card executes a block of instructions only when the specified condition is satisfied. The (optional) else card provides an alternative branch to be executed when the condition of the last if statement is not satisfied.

Fig. 5. Example of a two-way selection statement: if path ahead then move forward, else turn left.

Indentation identifies the instructions belonging to the THEN and ELSE branches. The same graphical convention used to delimit loop bodies is therefore reused to represent alternative execution paths.

The example in Fig. 5 shows a simple navigation strategy in which Roby moves forward when a free cell is available and turns left otherwise. Although simple, this example illustrates how program execution can depend on environmental conditions rather than on a predetermined sequence of instructions.

3.5 Expressiveness

Fig. 6. The *universal move*: a path-following program that iteratively executes a context-aware navigation strategy until the target is reached.

By combining sequence, repetition, selection, and nesting, CodyRoby supports the construction of programs whose behavior adapts to the environment during execution.

Figure 6 presents a path-following program that combines goal-oriented repetition with nested conditional statements to guide Roby toward the target. Rather than encoding a specific route, it specifies a navigation strategy (called *universal move*) that is repeatedly applied until the target is reached.

This example demonstrates how a small set of primitive instructions and control structures can generate behaviors that are not tied to a particular trajectory but emerge from the interaction between the program and the environment. It also illustrates the expressive power achieved through the combination of the fundamental constructs of structured programming within a compact and embodied language.

4 Embodied Execution

In a typical programming environment, program execution takes place inside a computer and most computational processes remain inaccessible to direct observation. In CodyRoby, execution is performed by a human or physical agent operating within a shared environment. As a consequence, many aspects of computation become directly observable.

The embodied nature of CodyRoby can be interpreted in the broader context of embodied and enactive approaches to cognition. Rather than treating cognition as the manipulation of abstract symbols detached from physical experience, these perspectives emphasize the role of bodily action and interaction with the environment in the construction of meaning and knowledge [16]. CodyRoby applies this principle to programming by representing computational structures through physical artifacts and by enacting program execution through observable actions performed within a shared environment.

Before discussing specific aspects of embodied execution, it is worth noting that CodyRoby makes explicit a fundamental characteristic of programmable systems: the separation between the space in which a program is represented and the space in which its effects take place. Programs are arranged as sequences of cards in a dedicated program space, while their effects are produced in the environment represented by the grid.

Execution requires Roby to progress simultaneously through these two spaces. On the one hand, Roby advances through the program one instruction at a time, effectively performing a fetch-and-execute cycle in the sense originally described by von Neumann [11]. On the other hand, Roby moves within the grid according to the effects of the currently executed instruction. The relationship between program execution and state evolution, which is normally hidden inside a processor, becomes directly observable through embodied action.

The following sections describe some of the computational processes that become visible through this execution model.

4.1 Visible State

The execution state is represented by the position and orientation of Roby within the grid. Unlike conventional programming environments, where state is typically encoded through variables and data structures, the CodyRoby state is represented through spatial relationships and physical artifacts.

Position and orientation are continuously visible throughout the execution and can be observed by all participants. State changes therefore become immediately perceptible and do not require symbolic interpretation.

4.2 Visible Control Flow

Control structures not only define program behavior but also shape the observable execution process.

Unlike elementary instructions, which directly modify the execution state in the environment, control structures act on the program space. Movement instructions affect Roby's position and orientation within the grid, thereby producing visible changes in the environment. Repetition and selection constructs, by contrast, determine how execution progresses through the program space, deciding which instructions should be executed, repeated, or skipped.

This distinction becomes directly observable during execution. While movement instructions produce visible changes in the trajectory, control structures produce visible changes in the execution flow. Participants can therefore observe not only the effects of computation, but also the mechanisms that govern the progression of the program itself.

The execution of a repetition construct requires repeatedly revisiting the loop body according to the condition specified by the Repeat card. In fixed-count loops, removable unit markers provide a visible and manipulable representation of the iterations still to be executed. At each iteration, one marker is removed from the card and the loop terminates when no markers remain. This mechanism externalizes the iteration count, allowing participants to execute count-controlled repetition without introducing variables or relying on counting skills. The state of the repetition construct therefore remains continuously observable throughout execution.

In condition-controlled and goal-oriented loops, execution repeatedly returns to the condition before entering the loop body again. Similarly, conditional constructs require the evaluation of a physical condition in the environment, before one of the available branches can be selected.

As a result, control flow is not merely represented in the program but becomes part of the observable execution process.

4.3 Visible Sensing and Context Awareness

Condition-controlled loops and conditional statements introduce sensing and context awareness into program execution.

Unlike fixed-count repetition, whose behavior is entirely determined by information encoded in the program, these constructs require information to be acquired from the environment during execution. Whether a cell is free, whether a target has been reached, or whether another condition is satisfied can only be determined at runtime.

In CodyRoby, this sensing activity becomes explicit and observable. Conditions do not represent passive properties of the environment; they represent instructions to inspect the environment. When a condition is encountered, Roby must actively determine whether it is satisfied by using the available sensory capabilities. The evaluation process therefore becomes part of the visible execution.

At the same time, sensing is controlled by the program itself. Roby does not continuously inspect the environment but performs the required checks only when instructed by a condition contained in the program. The program therefore determines not only which actions are executed, but also when information should be acquired from the environment.

As a consequence, context-aware behavior emerges from the interaction between program and environment. Participants can observe when sensing occurs, how conditions are evaluated, and how the resulting information influences the subsequent execution of the program.

4.4 Multiple Perspectives on Computation

Because execution takes place in a shared physical environment, computational processes become accessible simultaneously to the performer and to observers.

Participants can follow the evolution of the state, observe the execution of instructions, inspect the effects of control structures, and discuss alternative behaviors while the computation unfolds. Unlike conventional programming environments, where execution is largely confined to the interaction between a user and a machine, CodyRoby naturally supports collective observation and participation.

This collective visibility enables a wide variety of collaborative, competitive, and game-based activities in which embodied programming constitutes the core interaction mechanism. Participants may cooperate in the design, execution, debugging, or optimization of programs, or engage in challenges that require predicting, constructing, or modifying behaviors under specific constraints. Programming therefore becomes not only an individual activity but also a social and communicative experience.

The physical nature of the language further supports this flexibility. Instruction cards can be manipulated individually, assembled into programs, or collected into decks. Decks can be shuffled and distributed among participants to introduce randomness and uncertainty into programming challenges and games. The resulting activities combine algorithmic reasoning with social interaction, strategic thinking, and playful exploration while preserving the underlying computational structure.

By making computational processes collectively observable, CodyRoby enables collaborative, competitive, and game-based activities that transform the exploration and communication of computational concepts into a shared engaging experience.

4.5 Debugging and Execution Verification

Unlike conventional programming environments, where the correctness of instruction execution is normally assumed because the CPU is trusted, CodyRoby distinguishes between programming errors and execution errors. A programming error for which Cody is responsible occurs whenever a program does not satisfy its specification, either because it fails to reach the intended destination or because its execution violates one or more constraints of the environment, such as leaving the grid, leaving a prescribed path, or entering a blocked cell. An execution error occurs whenever the participant playing the role of Roby misunderstands or incorrectly executes an instruction, causing the observed execution to diverge from the behavior prescribed by the program.

Because CodyRoby activities are typically performed in groups and both the execution state and the progression of control flow remain continuously observable, participants can collectively detect, localize, and discuss both programming errors and execution errors.

To this purpose, both the grid and the program itself must remain visible to all participants throughout the activity. Each execution step is coordinated by Cody, or by another participant who makes the fetch-and-execute cycle explicit by identifying the next instruction, reading it aloud, and updating iteration markers of count-controlled loops whenever applicable.

The group therefore acts as a tester, verifying both the correctness of the program and the correctness of its execution. As a consequence, debugging and execution verification become inherent components of CodyRoby activities. Rather than being reserved for specialized testing phases, they naturally emerge from collective observation and discussion, making verification an integral part of the embodied programming experience.

5 Conclusions

This paper presented CodyRoby as an embodied programming environment for computational thinking. We described its environment, language, and execution model, showing how sequence, selection, iteration, and nesting are represented through a small set of coherent visual constructs.

Unlike conventional programming environments, CodyRoby externalizes key aspects of computation through tangible artifacts and embodied actions. Program state, control flow, sensing activities, and execution processes become directly observable and can be collectively inspected, discussed, and manipulated during execution. By making visible the relationship between program space and environment space, CodyRoby exposes computational mechanisms that are normally hidden inside a machine. In this sense, it externalizes the notional machine underlying program execution, making it embodied, collectively observable, and directly manipulable.

The paper focused on the description of the language and its execution model rather than on the educational activities enabled by the environment, teachers' experiences, or learning outcomes. Future work will investigate these aspects through classroom studies and empirical evaluation. By documenting the design principles and computational structures underlying CodyRoby, we aim to provide a foundation for future research on embodied programming, computational thinking, Computer Science education, and the communication of computational ideas.

Acknowledgments. The author would like to thank the thousands of passionate teachers who have brought CodyRoby into their classrooms over the last twelve years. Their experience, creativity, and feedback have played a crucial role in the refinement, dissemination, and continuous evolution of the environment.

References

1. Bell, T., Alexander, J., Freeman, I., Grimley, M.: Computer science unplugged: school students doing real computing without computers. New Zealand J. Appl. Comput. Inf. Technol. **13**(1), 20–29 (2009)
2. Bell, T., Lodi, M.: Constructing computational thinking without using computers. Construct. Found. **14**(3), 342–351 (2019)
3. Bell, T., Witten, I.H., Fellows, M.: Computer Science Unplugged. University of Canterbury, Christchurch (1998)
4. Bogliolo, A.: Vi presento Cody e Roby. https://www.codeweek.it/vi-presento-cody-e-roby/. Accessed 26 July 2026
5. Bogliolo, A.: CodyRoby EN (2016). https://codemooc.org/codyroby/EN/. Accessed 26 July 2026
6. du Boulay, B., O'Shea, T., Monk, J.: The black box inside the glass box: presenting computing concepts to novices. Int. J. Man Mach. Stud. **14**(3), 237–249 (1981)
7. Chen, P., Yang, D., Metwally, A.H.S., Lavonen, J., Wang, X.: Fostering computational thinking through unplugged activities: a systematic literature review and meta-analysis. Int. J. STEM Educ. **10**(1), 47 (2023)
8. Google: Build with Blockly (2014). https://docs.blockly.com/. Accessed 26 July 2026. Moved from the original project page https://code.google.com/archive/p/blockly/
9. Kwon, K., Brush, T.A., Kim, K., Seo, M.: Embodied learning for computational thinking in a mixed-reality context. J. Educ. Comput. Res. **62**(8), 1939–1960 (2025)
10. Lindgren, R., Johnson-Glenberg, M.: Emboldened by embodiment: six preconditions for research on embodied learning and mixed reality. Educ. Res. **42**(8), 445–452 (2013)
11. von Neumann, J.: First Draft of a Report on the EDVAC. University of Pennsylvania, Philadelphia (1945)
12. Papert, S.: Mindstorms: Children, Computers, and Powerful Ideas. Basic Books, New York (1980)
13. Papert, S.: Turtle geometry: A mathematics made for learning. In: Mindstorms. Children, Computers and Powerful Ideas, pp. 55–94 (1980)
14. Pattis, R.E.: Karel the Robot: A Gentle Introduction to the Art of Programming. John Wiley & Sons, New York (1981)
15. Resnick, M., et al.: Scratch: programming for all. Commun. ACM **52**(11), 60–67 (2009)
16. Varela, F.J., Thompson, E., Rosch, E.: The Embodied Mind: Cognitive Science and Human Experience. MIT Press, Cambridge (1991)
17. Weintrop, D.: Block-based programming in computer science education. Commun. ACM **62**(8), 22–25 (2019)
18. Wilson, M.: Six views of embodied cognition. Psychon. Bull. Rev. **9**(4), 625–636 (2002)
19. Wing, J.M.: Computational thinking. Commun. ACM **49**(3), 33–35 (2006)

Making CS Education More Accessible Through Creative Robot Personalisation

Laura Cesaro(✉), Anna Polato, and Emanuele Menegatti

Department of Information Engineering, University of Padua, Padua, Italy
{laura.cesaro,anna.polato}@phd.unipd.it, emanuele.menegatti@unipd.it

Abstract. Educational robotics offers a tangible, hands-on entry point into computer science (CS), but fostering inclusive participation remains a challenge under everyday classroom constraints: limited curricular time, heterogeneous student backgrounds, and limited opportunities for creative design when using pre-assembled robots. This work investigates creative robot personalization as a practical pedagogical strategy to introduce creative agency and broaden participation in middle school CS education. During a 10-h course conducted during regular school hours, 54 lower secondary school students aged 12–13 worked in pairs to customize ready-to-use robots before engaging in block-based programming. We analyzed robot descriptions collected before and after the programming activities, questionnaire responses, and technical assessment scores. Results indicate that personalization provided an expressive entry point: after the programming activities, students' robot descriptions shifted from imaginative, anthropomorphic narratives to technical accounts, with female-female pairs producing particularly rich and detailed final descriptions.

Keywords: Educational robotics · CS Education in Middle school · Inclusive learning

1 Introduction

Computer science education plays an increasingly important role in preparing students to understand, shape, and participate in a digital society [6,7,13]. However, access to computing is not only a matter of curriculum: students also need to perceive computer science as a field in which they can belong, contribute, and express ideas. This is particularly relevant in middle school, when interests and self-efficacy beliefs toward technical domains begin to consolidate, and girls often show lower interest than boys in computer science and robotics [5,11,18].

Educational robotics is often proposed as a promising way to introduce programming and computational thinking through tangible hands-on activities [10]. However, bringing robotics into ordinary classrooms, where participation is not self-selected, raises practical and pedagogical challenges. These constraints raise

M. Löffler et al. (Eds.): CMSC 2026, LNCS 16951, pp. 87–102, 2027.
https://doi.org/10.1007/978-3-032-38887-2_7

a central design question: how can introductory robotics activities be structured to enable broader participation, particularly among students who may not initially identify with technical domains?

One strategy to address this challenge is to design learning experiences that integrate creativity, personal expression, and agency. In educational robotics, hands-on and age-appropriate activities have been shown to support girls' interest in technology and engineering [15]. This perspective aligns with the constructionist tradition [9,14], which emphasizes that students learn most effectively when actively engaged in building and personalizing meaningful artifacts. Personalization and storytelling in particular have been shown to enhance motivation and the sense of ownership over learning [1]. In addition, the appearance of the robot also plays a significant role in influencing the attitudes and opinions of female and male users [3]. This paper reports on a classroom experience in which robot personalization was used as a strategy to foster inclusive participation in a 10-hour robotics and programming course for lower secondary school students. We investigate three research questions:

- **RQ1**: How do students perceive and engage with creative robot personalization in an introductory robotics activity?
- **RQ2:** How do students' descriptions of robots evolve after programming activities?
- **RQ3:** What connections emerge between robot personalization, student engagement, and technical learning?

Given the focus on inclusion, we also consider whether patterns differ across gender group compositions where relevant.

2 Related Works

Gender, Inclusion, and Participation in CS/STEM Education. Gender disparities in STEM, particularly in computer science, are shaped by interests, self-efficacy beliefs, stereotypes, and perceptions of belonging [5,18]. Gender-interest stereotypes in computer science and engineering emerge early and can negatively affect girls' motivation, belonging, and participation [11], making middle school a particularly relevant context [17]. Recent work also highlight that computational thinking projects in secondary education may support students' attitudes toward STEM and future educational choices [8]. Together, these studies suggest that inclusive CS education should not only provide access to technical content, but also create learning contexts in which diverse students can develop confidence, interest, and a sense of belonging.

Educational Robotics and Inclusive Engagement. In educational robotics, gender inclusion has been examined along several dimensions: Maigaard et al. [10] synthesize studies suggesting that girls' engagement in robotics can be influenced by the choice of technology, the way activities are contextualized, teaching approaches, and team composition. The same review also highlights the role of contextualization: girls often respond more positively to robotics activities framed around social, human-centered, or care-oriented applications than to purely task- or function- oriented challenges. Collaborative design is another relevant dimension of inclusive STEM activities. McLean et al. [12] show that activity design can support the development of engineering identity, suggesting that collaborative rather than competitive settings may shape inclusive participation. Beyond the structure of the activity, the robot itself can shape how students interpret and engage with robotics: educational robots can be associated with gendered perceptions [3], and studies with primary school students suggest that human-like and biomimetic designs show broader cross-gender appeal than mechanical designs [4].

Personalization and Creative Agency as a Pedagogical Approach. The value of creative agency and personalization in learning is central to constructionist theory, which frames learning as most effective when students actively create meaningful artifacts [14] and recognizes concrete and relational approaches as legitimate ways of engaging with computational ideas [16]. Building on these perspectives, designing, personalizing, and sharing computational creations has been emphasized as central to learning [9], and maker-oriented approaches have shown that integrating creative expression with technical making can increase engagement and broaden participation [1,2,8]. In educational robotics, such approaches suggest the potential of treating robots not only as technical devices, but also as objects that students can personalize and make meaningful.

Positioning of this Work. While these studies provide useful principles for designing inclusive robotics activities, less attention has been paid to how such principles can be applied under ordinary classroom constraints: limited time, non-self-selected participation, and ready-to-use robots. This experience report examines robot personalization as a simple strategy to introduce creative agency into such settings, focusing on how lower secondary school students engaged with personalization before programming and how this related to their engagement, robot descriptions, and technical learning.

3 Methodology

Context and Participants. The study was conducted in a lower secondary school in Padova, Italy, involving three classes during regular school hours. A total of 54 students participated (27 female and 27 male, 12–13 years old). All students in the three classes were included; participation was not voluntary or self-selected, which distinguishes this study from much prior work on gender inclusion in educational robotics, which has often been conducted in extracurricular or voluntary settings. Students worked in pairs. Due to the unbalanced gender composition of each class, most pairs were same-gender (10 female-female, FF; 8 male-male, MM), with the remaining 9 pairs being mixed-gender (MF).

Learning Activity Design. Each class participated in a 10-h robotics and programming course, structured as five 2-hour sessions held once a week. The program was designed and facilitated by the researchers, who are also authors of this paper. Each pair worked with one Makeblock mBot2, a ready-to-use robot already available at the school, and programmed it using mBlock, its dedicated visual block-based programming environment. Because the pre-assembled nature of the robot limits traditional construction opportunities, the curriculum started with a personalization phase to spark creativity and engagement before the programming activities. Each pair received an identical cardboard template to customize through cutting and decoration, a process through which many students gave their robot not only a distinctive appearance, but also an emerging identity. Before introducing technical concepts, students presented these customized creations to the class. Although students were not explicitly instructed to create a story, many pairs spontaneously described their robots' features, future capabilities, and even their "feelings." This approach integrated key design principles such as learner agency, collaborative group work, scaffolded instruction, personalized learning, and storytelling, with the aim of supporting engagement, conceptual understanding, and active, participatory learning. Following this initial presentation, the curriculum shifted toward programming and robot functionality, starting with basic movement and rotation commands through a series of incremental mini-tasks. Students then explored sensor-based behaviors, using an ultrasonic distance sensor for obstacle avoidance and an RGB sensor for color recognition and line-following activities. The technical sequence concluded with activities involving built-in light and sound sensors and the integrated gyroscope. During the final lesson, the course returned to the robot personalization activity. Each pair was asked to describe their robot once again, reflecting on its features and capabilities after the programming activities. Students also completed a final questionnaire.

Data Collection. Data were collected through multiple instruments. *Before the course,* each pair presented their customized robot to the class. These oral presentations were transcribed by the researchers facilitating the course to capture students' thoughts, creative choices, and storytelling elements. *At the end of the course,* students completed two online forms:

- a pair-level form in which each pair provided a written description of their robot, detailing its design, features, and capabilities.
- an individual form, divided into two parts:
  - (a) semi-structured questions about students' perceptions of robot customization, focusing on creativity, engagement, and the role of personalization in the learning experience;
  - (b) a set of 10 closed-ended technical questions assessing understanding of robot components and programming concepts covered during the course.

Data Analysis. This study combines qualitative analysis of students' robot descriptions with descriptive quantitative analysis of questionnaire responses and technical assessment scores. The robot descriptions collected both at the start and at the end of the program were analyzed qualitatively through an inductive approach by the research team, allowing categories to emerge directly from the data. Both initial and final descriptions were examined with respect to four aspects: the names given to the robots, the characteristics attributed to them, the gender assigned, and the explicit or implicit mention of programming. The final descriptions were additionally classified on a 5-point scale to assess the degree to which students referred to the robot's actual capabilities. Coding was performed independently by two authors; cases of disagreement or uncertain classification were discussed among the research team until agreement was reached. Questionnaire responses were analyzed descriptively. Given the experience-report nature of the study and the small sample size, quantitative results were interpreted descriptively rather than inferentially. Technical understanding was assessed through 10 closed-ended questions, assigning 2 points per correct answer for a maximum score of 20 points.

4 Findings

4.1 Analysis of Robot Descriptions

- Robot names: What names were given to the robots?
- Robot characteristics: What characteristics were attributed to the robots before and after the course?
- Assigned gender: Was a gender assigned to the robots before and after the course?
- Functioning of the robot: Did the descriptions indicate that the robot requires programming to work?

Robot Names. 13 pairs out of 27 gave their robot a human name, often inspired by the designers' own names or nicknames. Four referenced natural elements, three pairs assigned names based on colours, and three chose names related to technology and mechanics, including one pair who simply named their robot 'Robottino' (the Italian word for 'Little Robot'). Two pairs created imaginative names, one selected names associated with sports, and one referenced a commercial product. Human-related names were the most common category across pair types (see Fig. 1), although some gender-related patterns emerged:

- MM pairs used a narrower range of names, mostly based on human-related and technology-related categories, with some names associated with products or sports.
- MF pairs showed a strong preference for human-related names and occasionally used nature-related names; other categories did not appear in this group.
- FF pairs used a broader range of name types, including technology-related, imaginative, and nature-inspired names.

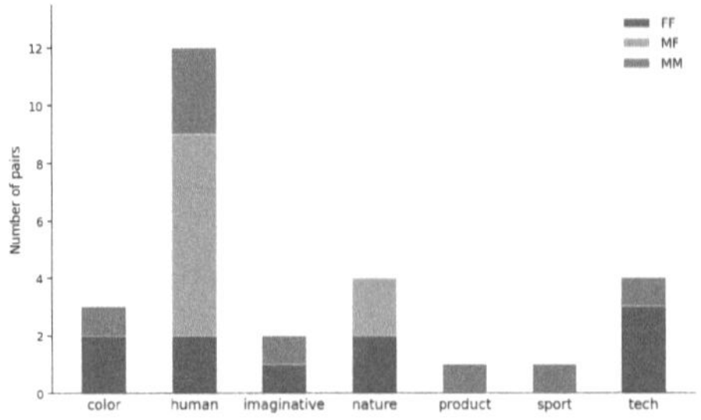

Fig. 1. Categories of names assigned to the robots by students.

Robot Characteristics . In the initial descriptions, collected after personalization and before programming activities, approximately half of the pairs used adjectives and qualities typically associated with humans. No pair listed specific technical characteristics of the robot, although some mentioned mechanical elements (e.g., wheels), astronautical concepts (e.g., "space robot"), or digital features (e.g., "always connected"). Seven pairs incorporated fantastical elements (e.g., dragon, ghost, mascot), four referenced elements of nature (e.g., rainbow, flower, turtle), and two focused primarily on aesthetic characteristics. In the storytelling, a strong connection often emerged between the robots and students' personal experiences, emotions, and interests. These initial descriptions suggest that personalization supported identity projection and creative engagement; representative examples are reported in Appendix A.

The final descriptions, collected after the programming activities, showed that most pairs referred to at least some of the robot's technical features. However, in some cases, a tendency toward imaginative elements and anthropomorphic traits persisted. The descriptions were classified on a 5-point scale: (1) not based on the robot's capabilities, (2) based on a single robot characteristic, (3) based on the robot's fundamental capabilities, (4) partially based on the robot's capabilities and partially humanized, (5) fully based on the robot's capabilities. The overall distribution by level and by pair type is shown in Fig. 2).

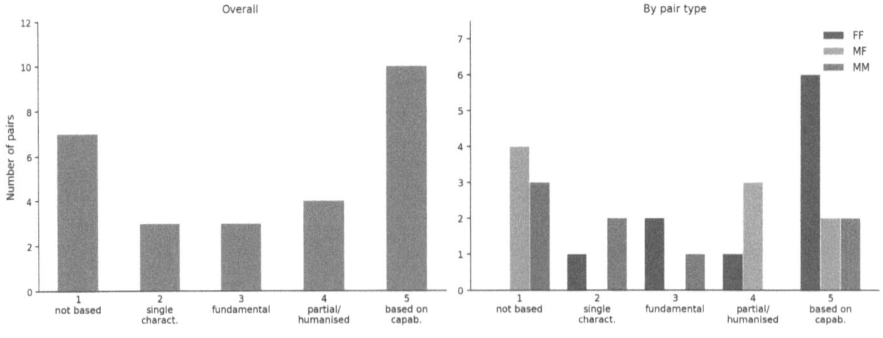

Fig. 2. Robot knowledge based on descriptions, overall, and by pair type.

At the end of activities, pairs included both descriptions of physical components (e.g. lights, display, buttons, sensors, actuators, motherboard) and specific functionalities (e.g., *"It can move forward and backward, emit sounds, display messages, avoid obstacles using sensors ..."*). FF pairs tended to provide more comprehensive and detailed explanations of how robots work; MM pairs showed intermediate results, with several specific descriptions but less consistent detail; MF pairs more often provided descriptions that were incomplete or only weakly connected to the robot's features and capabilities.

Assigned Gender. Before the activities, male gender attribution predominated (19 cases), likely influenced by the grammatical gender of the word "robot" in Italian. In 6 cases, no gender was specified, while in 3 cases, the robot was described as female. Figure 3 shows the distribution before and after the course.

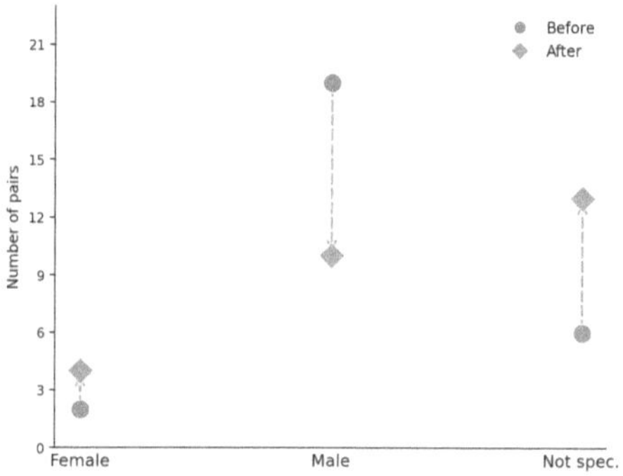

Fig. 3. Gender given to the robots before and after activities.

In the final descriptions, a noticeable shift emerged toward not assigning a specific gender to the robot (13 cases), suggesting that students increasingly perceived the robot more as a technological object rather than as a projection of personal or human characteristics. The number of robots described as female slightly increased (from 3 to 4 cases).

Functioning of the Robot. In analyzing the student-provided descriptions, we aimed to assess their understanding of programming as a fundamental aspect of the robot's functionality. Responses were categorized based on the presence or absence of programming references: absence, implicit reference, and explicit reference. Results are shown in Fig. 4.

Although more than half of the pairs explicitly acknowledged programming or referred to it using non-technical terms, more than 40% did not mention the necessity of programming for the robot to perform tasks. Most FF pairs referred to programming indirectly, most MF pairs did not mention programming, MM pairs mentioned programming more frequently than the other groups.

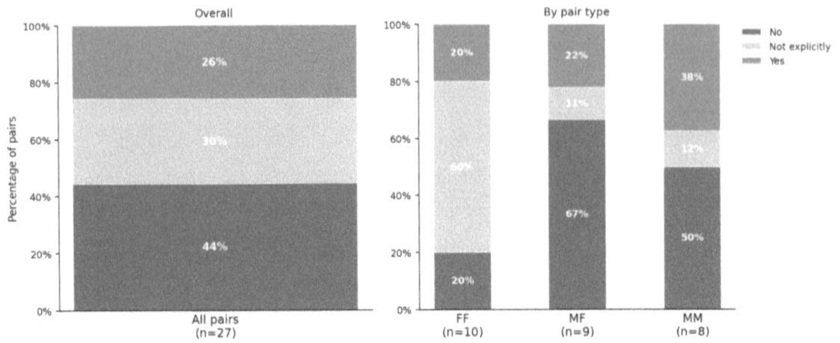

Fig. 4. Programming references in final robot descriptions, overall (left) and by pair type (right).

4.2 Questionnaire Analysis

A total of 54 students completed the individual form. Although the class registers indicated an equal gender distribution, with 27 male and 27 female students, 8 students chose not to specify their gender in the individual form. As a result, the data were analyzed across three groups: 25 male students, 21 female students, and 8 students whose gender was not specified. The form included two sections: one on students' perceptions of personalization and programming, and one with 10 closed-ended technical questions on robot programming and functionality.

Student perceptions about robotics activities

Item 1. *"How important was it for you to customize the robot?"* Students rated this item on a Likert scale from 1 (Not at all important) to 5 (Extremely important). Results are shown in Fig. 5. Responses were generally moderate across gender groups: most female responses were neutral or somewhat positive, while male responses were more distributed, with 2 as the most frequent value but also several positive ratings. Mean scores were close across groups: 2.86 for females, 3.04 for males, and 3.0 for students with non-specified gender. Open-ended explanations were grouped according to whether personalization was perceived as use-

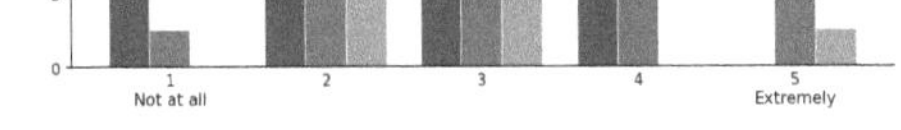

Fig. 5. Perceived importance of customization, by gender.

ful or less useful/accessory; representative responses are reported in Appendix A.

Item 2. *"Do you think programming has helped you better understand how the robot works?"*—Fig. 6(a). Most students (51 out of 54) reported that programming helped them understand how the robot worked. The perception of the usefulness of programming was strongly positive across gender groups.

Item 3. *"Do you think that robot customisation has made robotics activities more interesting?"*—Results are shown Fig. 6(b). Customization was perceived as making the activities more interesting by 30 students out of 54. This perception was slightly more common among male students, while female students' responses were more balanced.

Item 4. *"In the final description of the robot, did you take into account what you learnt?"* Fig. 6(c). Most students (45 out of 54) indicated that they applied what they had learned to the final description of their robot. Overall, students' explanations suggest that most of them connected the final description to the learning process in different ways, including by referring to programming, theoretical knowledge, new terminology, and changes in their initial expectations. Representative student explanations are reported in Appendix A.

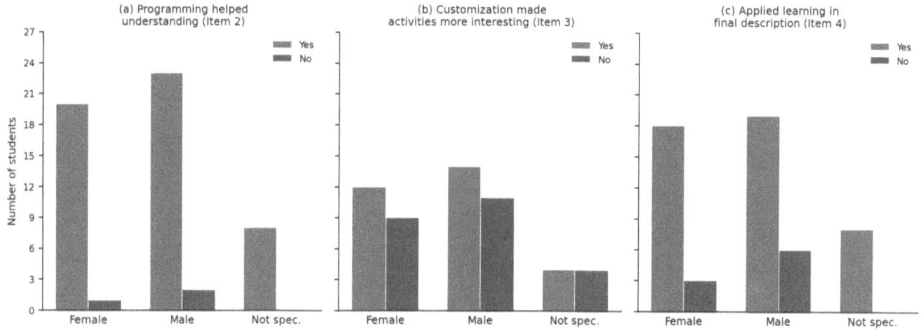

Fig. 6. Students' perceptions of the robotics activities: programming helpfulness (a), interest added by customization (b), use of learning in final description (c).

Student Knowledge About Robotics and Programming. The second part of the questionnaire assessed students' technical understanding of robotics and programming. For each correct answer students were awarded 2 points, for a maximum possible score of 20 points. Results are shown in Fig. 7. Most students scored in the middle to higher range (12–18 points), with over 60% scoring above 12 points. This suggests that the robotics activities were relatively effective in promoting technical learning for most students. There does not appear to be a clear gender-based gap in the scores overall, with an average score of 12.48 ±

4.09 for female students and 13.76 ± 3.57 for male students. Both groups had the same median value of 14. The students who did not specify their gender showed a varied score distribution, with an average score of 10 ± 5.86 and a median of 10. Figure 7 shows how technical scores were distributed between the

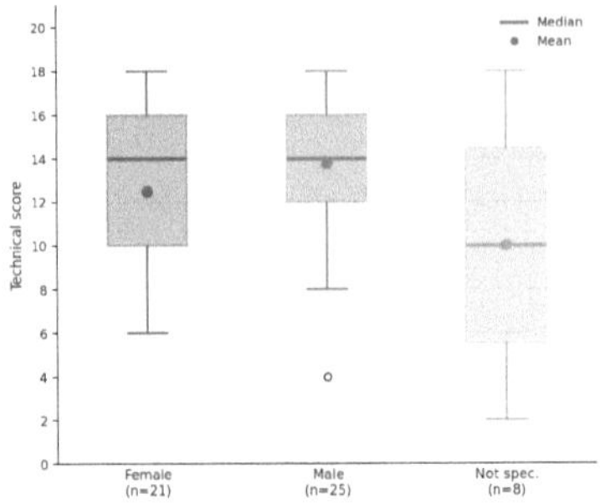

Fig. 7. Technical score distribution by gender.

perceived importance of robot personalization within each gender group. The figure does not show a clear linear relationship between the perceived importance of personalization and technical scores, suggesting that personalization is better interpreted as an engagement-related element than as a direct predictor of technical learning outcomes.

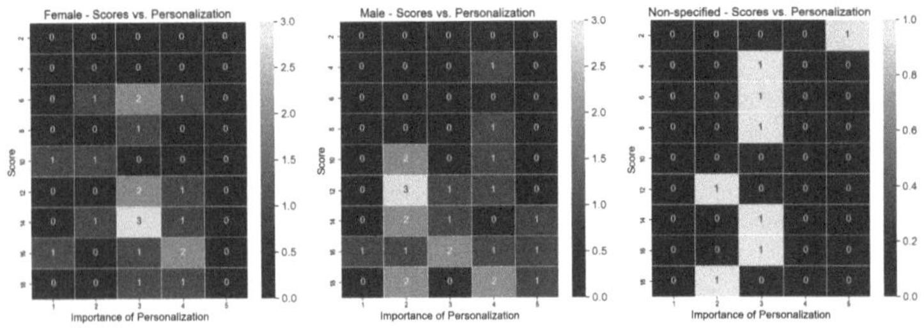

Fig. 8. Technical scores and perceived importance of personalization, by gender.

5 Discussion

The findings of this classroom experience offer a nuanced picture of how creative robot personalization functioned as a pedagogical strategy during regular school hours. We discuss the results in relation to each research question, considering gender group composition where relevant.

RQ1: *How do students perceive and engage with robot personalization in an introductory robotics activity?* Students' responses suggest that robot personalization was perceived as a moderately important component of the activity. Students who valued personalization highlighted the sense of ownership and creative involvement it provided, while those who rated it less positively tended to distinguish between the aesthetic dimension of the robot and its functional capabilities, expressing greater interest in the programming activities. Taken together, these findings suggest that, overall, personalization appeared to work primarily as an expressive and creative entry point, although not all students perceived it as directly connected to technical learning.

RQ2: *How do students' descriptions of robots evolve after programming activities?* The analysis of robot descriptions before and after the programming activities shows an evolution in how students conceptualized their robots. Initial descriptions were largely narrative and anthropomorphic, with students projecting emotions, personal experiences, and human qualities onto their robots. This strong tendency toward humanization in the initial descriptions can be interpreted through a constructionist lens, as students connected the activity to their own identities and experiences. By the end of the program, most pairs demonstrated at least partial understanding of the robot's technical features, with a general shift toward more functional descriptions. However, the degree of this shift varied notably across gender groups. FF pairs produced the most technically detailed final descriptions. MM pairs showed a more varied distribution, while MF pairs remained more anchored to narrative elements.

RQ3: *What connections emerge among robot personalization, student engagement, and technical learning?* The relationship between personalization, engagement, and technical learning appeared more complex than a simple linear connection. Programming was nearly unanimously perceived as useful for understanding the robot, and technical scores showed no significant gender gap. A more nuanced pattern emerged when comparing how different groups explained their technical understanding. FF pairs produced the most technically detailed final descriptions, yet their references to programming were predominantly implicit, conveyed through functional descriptions of what the robot can do rather than explicit technical vocabulary. This contrasts with MM pairs, who more frequently used explicit programming references but showed greater variability in the quality of their final descriptions. This suggests that FF pairs may have developed a different pathway to technical understanding, one that integrates functional knowledge within a more narrative and relational register. These findings tentatively suggest that the combination of personalization and programming may have been particularly generative for FF pairs, who showed both rich creative engagement in the initial descriptions and the most technically detailed final descriptions. Whether this reflects a specific synergy between creative agency and technical learning for this group, or other factors related to pair dynamics, remains an open question that warrants further investigation.

Limitations: This study has several limitations that should be acknowledged. First, the relatively small number of participants in each gender group limited the robustness of gender-based comparisons. Second, no control group was included, as all classes in the school participated in the intervention. Finally, the grammatical gender of the word "robot" in Italian may have influenced students' initial gender attributions.

6 Conclusions

This paper described a classroom intervention in which robot personalization was used as an initial strategy to engage lower secondary school students in programming and computational thinking during regular school hours. The findings suggest that personalization can offer a valuable creative entry point into educational robotics. Programming was positively evaluated by most students, and technical learning outcomes appeared broadly comparable across gender groups. A notable pattern emerged in the final robot descriptions: female-female pairs produced particularly rich descriptions, raising the possibility that the combination of creative engagement and technical instruction may have opened different pathways to learning. These findings suggest that inclusive robotics curricula should not assume that personalization will engage girls specifically, but rather provide multiple entry points: creative, social, and technical, through which different students can connect with programming and robotics.

Generative AI Disclosure. The authors used an AI-assisted editing tool to improve grammar, clarity, and readability during manuscript preparation.

A Additional Material

This appendix presents supplementary qualitative data and materials from the classroom study. It includes a visual example of robots personalization (Fig. 8), direct excerpts of the students' robot descriptions, responses regarding the perceived importance of customization, and student reflections on the knowledge applied during the curriculum (Fig. 9).

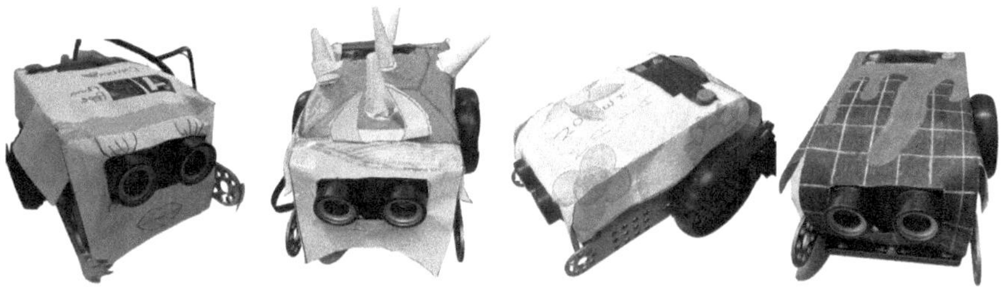

Fig. 9. Physical appearance of robots as customized by students using the cardboard template.

A.1 Initial robot descriptions

During the initial presentation phase, a strong connection emerged between the robots and students' personal experiences, emotions, and interests across pair types:

1. *"He likes playing soccer (MM), he's intelligent but loves to party, he gets angry (MF), he enjoys eating hamburgers (MM), he doesn't like school (MM), he does well in every subject except history and Italian (MF)..."*
2. *"He's always happy (MF), he's very calm (MF), he's wise (MF), he's always smiling (FF), spreads joy (FF), she is sad because she came from the sky and can no longer fly on Earth (FF)..."*
3. *"He's named after a famous singer (FF), he is inspired by Super Mario (MM), he trains three times a week (MF)..."*

On two occasions, classmates drew parallels between the robot's attributed behaviour and the presenter: *"...he never gets angry, just like you, D.!"*; *"...he trains for soccer three times a week—are you talking about yourself, F.?"*

One FF pair initially perceived the robot as a wild and uncontrollable entity: *"It's a wild robot that needs to be tamed because we don't know how to manage it."*

A.2 Perceived importance of customization (Item 1)

Students were asked to rate the importance of robot customization and to explain their rating in an open-ended field. Tables 1 and 2 report representative responses, grouped by whether personalization was perceived as useful or not, and by gender.

Table 1. Responses and explanations given to Q1 – Personalization perceived as useful.

Females	Males	Not Specified
It felt like I was participating, at least partially, in creating the final parts of the robot, and I felt involved	It was fun and added an artistic side to programming	It was great to make clothes for it
Aesthetic is important, and it makes the robot unique.	Useful to distinguish it from others	It's a creative activity
Allowed us to describe the group's personalities	I like personalizing things	Made the robot more beautiful
To give the robot your personality	I like my robot is personalized	
Not essential but nice (3)	To make it different	
It was fun (4)	It was interesting	
I like this activity (2)	I like it (2)	
It was interesting (2)		

Table 2. Responses and explanations given to Q1 – Personalization perceived as not or less useful.

Females	Males	Not Specified
It became prettier, but its functions didn't change	The programming mode didn't change	It was better without the personalization
I'm interested in how to use the robot (2)	It would have been better without it	Not so interesting
Not useful	The outfit was uncomfortable	
	It was ugly	
	I didn't understand the purpose	
	Not very interesting	
	The robot was already nice	
	It became prettier, but its functions didn't change (2)	
	I didn't need it much	
	I was more interested in learning how to use it (2)	
	Not useful or interesting (3)	
	Fun but not useful (2)	

A.3 Knowledge applied in the final robot description (Item 4)

Students were asked whether they had taken into account what they learned when writing the final robot description, and to explain their answer. The following responses are grouped by positive or negative response.

Responses supporting "Yes":

- *"I considered what I learned because, in some way, it felt like the robot was doing those actions voluntarily and on its own, even though I was the one who programmed it."* (F)
- *"What I knew before wasn't even a third of what I was taught. I'm really happy with what I learned because it's very useful."* (F)
- *"In the final description, it said to include everything I knew. So at first, I just described the robot a little, but then I also thought about the theory and*

the programming. I tried to describe everything I could remember, as if it was for someone who didn't know anything about the robot." (F)
- *"At first, I thought it would be boring and hard to do, but in the end, it was fun—difficult, but not too hard."* (F)
- *"Because I understood new things."* (M)
- *"Because I used terms I didn't know before."* (M)
- *"Because it is important to remember these things."* (M)

Responses supporting "No":

- *"Because I won't use it in the future."* (F)
- *"Because I got caught up in describing it in a fun way."* (M)

References

1. Blikstein, P.: Maker movement in education: history and prospects. In: de Vries, M. (ed.) Handbook of Technology Education. Springer, Heidelberg (2018)
2. Buechley, L., Peppler, K., Eisenberg, M.: Textile messages: dispatches from the world of E-textiles and education. In: New Literacies and Digital Epistemologies (2013)
3. Cesaro, L., Franceschini, A., Badaloni, S., Menegatti, E., Rodà, A.: Gender biases in robots for education. In: Proceedings of BEWARE 2024. CEUR Workshop Proceedings, vol. 3881, paper 4 (2024)
4. Cesaro, L., Menegatti, E.: Robot design and inclusive practices: a pilot study on gender equity in STEM. In: Staffa, M., et al. (eds.) Social Robotics + AI. ICSR+AI 2025. LNCS, vol. 16133. Springer, Singapore (2026). https://doi.org/10.1007/978-981-95-2398-6_45
5. Cheryan, S., Ziegler, S.A., Montoya, A.K., Jiang, L.: Why are some STEM fields more gender balanced than others? Psychol. Bull. **143**(1), 1–35 (2017)
6. Cosgrove, J., Cachia, R.: DigComp 3.0: European digital competence framework, Publications Office of the European Union. Technical report (2025). https://doi.org/10.2760/0001149
7. European Commission: A STEM education strategic plan: Skills for competitiveness and innovation. Technical report, European Commission (2025). https://education.ec.europa.eu/sites/default/files/2025-03/STEM-Communication-graphic-version-updated_0.pdf
8. González-Gallego, S., Hernández-Pérez, M., Alonso-Sánchez, J.A., Hernández-Castellano, P.M., Quevedo-Gutiérrez, E.G.: A critical examination of the underlying causes of the gender gap in STEM and the influence of computational thinking projects applied in secondary school on STEM higher education. Front. Educ. **10**, 1537040 (2025). https://doi.org/10.3389/feduc.2025.1537040
9. Kafai, Y.B., Resnick, M.: Constructionism in Practice: Designing, Thinking, and Learning in A Digital World. Routledge, Abingdon (1996)
10. Maigaard, B.K.M.K., Weigelin, B.C., Larsen, J.C., Nielsen, J.: Using educational robotics to foster girls' interest in STEM: a systematic review. In: 2021 30th IEEE International Conference on Robot & Human Interactive Communication (RO-MAN), pp. 865–872 (2021)

11. Master, A., Meltzoff, A.N., Cheryan, S.: Gender stereotypes about interests start early and cause gender disparities in computer science and engineering. Proc. Natl. Acad. Sci. **118**(48), e2100030118 (2021). https://doi.org/10.1073/pnas.2100030118
12. McLean, M., Nation, J.M., Spina, A., Susko, T., Harlow, D., Bianchini, J.: The importance of collaborative design for narrowing the gender gap in engineering: an analysis of engineering identity development in elementary students. J. Pre-Coll. Eng. Educ. Res. (J-PEER) **10**(2), 2 (2020). https://doi.org/10.7771/2157-9288.1210
13. OECD: Education at a Glance 2023: OECD Indicators. OECD Publishing, Paris (2023). https://doi.org/10.1787/e13bef63-en
14. Papert, S.: Mindstorms: Children, Computers and Powerful Ideas. Basic Books (1980)
15. Sullivan, A., Bers, M.U.: Investigating the use of robotics to increase girls' interest in engineering during early elementary school. Int. J. Technol. Des. Educ. **29**(5), 1033–1051 (2019)
16. Turkle, S., Papert, S.: Epistemological pluralism and the revaluation of the concrete. In: Harel, I., Papert, S. (eds.) Constructionism, pp. 161–191. Ablex Publishing (1991)
17. Wang, D., Wang, G.: Improving career readiness in middle school students: a systematic review of intervention approaches. Front. Psychol. **16**, 1582195 (2025). https://doi.org/10.3389/fpsyg.2025.1582195
18. Wang, M.T., Degol, J.L.: Gender gap in science, technology, engineering, and mathematics (STEM): current knowledge, implications for practice, policy, and future directions. Educ. Psychol. Rev. **29**(1), 119–140 (2017)

From Pantomime to Bitomime – A CS Unplugged Approach to Binary Encoding and Decoding in Primary Education

Sonja Klein(✉), Lukas Lehner, Klaus Nigsch, and René Röpke

TU Wien, Favoritenstraße 9-11, 1040 Vienna, Austria
{sonja.klein,lukas.lehner,klaus.nigsch,rene.roepke}@tuwien.ac.at

Abstract. Encoding and decoding information are essential processes in computing. However, learning about them in primary school can be a challenge, due to their high level of abstraction. We present an unplugged workshop designed to introduce students aged 7–10 to binary encoding and decoding through a progressive abstraction approach. The workshop relies on embodied interaction, tactile and symbolic codes, as well as code verification exercises to guide students from intuitive, iconic representations toward abstract binary codes. Over the course of a year, the workshop was implemented in 20 individual sessions with a total of 422 students, offered in the context of our university's informatics outreach program. Based on repeated implementation and iterative refinement, we report practical observations and didactical insights suggesting that our embodied and tactile activities support younger students in developing an understanding of encoding and decoding as abstract processes. At the same time, we observed that the transition to binary representations remains a conceptual challenge requiring explicit scaffolding and reflection.

Keywords: Computer Science Unplugged · Binary Encoding · Primary School Education

1 Introduction

Foundational computer science (CS) concepts, such as data representation, encoding/decoding, and binary numbers, are widely recognized as core components of K–12 CS education [7,8,10,12]. Primary school represents a significant entry point for introducing such concepts, as early encounters with symbolic representation, sequencing, and abstraction can support children's later understanding of digital technologies [4,12,15]. However, in the Austrian primary school curriculum, CS is framed as an overarching, cross-curricular theme rather than as a dedicated subject, which may limit systematic and sustained learning opportunities for foundational CS concepts [5]. Meanwhile, understanding technological principles and fostering numerical reasoning through playful interaction is emphasized in subjects, such as math [5]. Computer Science Unplugged

M. Löffler et al. (Eds.): CMSC 2026, LNCS 16951, pp. 103–114, 2027.
https://doi.org/10.1007/978-3-032-38887-2_8

(CS Unplugged) provides a well-established approach to introducing computational concepts without digital devices, using tangible, playful learning activities grounded in constructivist learning principles [1,2]. By grounding abstract CS concepts in embodied and game-based activities, unplugged approaches support age-appropriate engagement with CS [1,9].

Building on this tradition, this paper presents an experience report on an unplugged workshop for primary school students (ages 7–10) that introduces binary en- and decoding through a progressive abstraction design. The workshop combines embodied interaction, tactile encoding, code verification, and symbolic encoding and decoding to guide students from intuitive iconic representations toward abstract binary codes. Based on 20 workshop implementations with a total of 422 students, this paper reports on the workshop design, practical observations, and didactical insights that emerged from repeated implementation.

2 Related Work

Since the introduction of CS Unplugged by Tim Bell et al. [3], binary encoding has been a recurring topic in unplugged CS education. Early activities such as *Count the Dots*[1] introduced binary representation through card-based tasks that focused on converting between decimal and binary numbers. Another activity – *Binary Weaving*[2] – relies on beads on a string to represent letters encoded using ASCII[3]. These activities established binary encoding as an accessible entry point for introducing symbolic representation and fostering computational thinking skills without the use of digital devices. More recent approaches have expanded binary encoding activities toward embodied and discovery-based learning formats, grounded in concrete non-computational topics. For example, Buttke and Krügel [6] used the *Cup Song*[4] as a basis for introducing symbolic and binary representations through movement-based encoding tasks. Their findings highlight that students actively moved between concrete and abstract representations during encoding and reflection phases, emphasizing the students' gradual learning process from clearly iconic codes to abstract binary systems. For younger students, Kordaki and Mpimpas [11] developed a binary encoding activity in which primary school students transformed pictures into binary grid representations. Their evaluation showed that most students were able to encode visual information using binary symbols, whereas more complex encoding tasks, such as encoding different colors through multi-digit codes, often required additional scaffolding.

Across these approaches, binary encoding and decoding activities often combine playful interaction, symbolic representation, and scaffolded abstraction.

1 https://classic.csunplugged.org/activities/binary-numbers/#count-the-dots, last accessed on 08-06-2026.
2 https://raft.net/idea_sheets/binary-weaving/, last accessed on 08-06-2026.
3 American Standard Code for Information Interchange.
4 A percussion song known by the movie *Pitch Perfect* by Jason Moore, released in 2012 (https://www.imdb.com/title/tt1981677/, last accessed on 08-05-2026).

However, existing approaches typically focus on isolated encoding and/ or decoding tasks. In contrast, the workshop presented in this paper combines embodied, tactile, and symbolic en- *and* decoding activities in a progressive-abstraction design that explicitly connects the encoding, decoding, transmission, and code verification of the alphabet within a single coherent workshop format.

3 Workshop Design

The workshop session is structured in seven steps: It begins with (1) an introduction to codes in everyday life, followed by (2) a charades-like pantomime game, (3) a haptic encoding activity, (4) a code verification activity, and (5) a word de- and encoding activity. It concludes with (6) *Bitomime*, a *binary* charades-like pantomime game, followed by (7) a joint reflection and discussion (see Table 1). While the workshop follows a sequential order, each activity can also be done individually. The sequential structure of knowledge moves from very concrete and symbolic codes towards abstract binary en- and decoding activities, which can be described as a semantic wave design proposed by Waite et al. [14], specifically for unplugged CS education for primary school students. In line with the theoretical grounding in constructivism of CS unplugged [1,3], the workshop follows a social-constructivist approach [13], using scaffolding, facilitator guidance, and peer interaction to support knowledge construction. Accordingly, we designed the activities so that students progressively construct, negotiate, and apply representational encoding and decoding conventions within the workshop session.

Table 1. Workshop Structure divided into seven steps and a break in between.

Step	Activity	Duration	Setting
1	Introduction	10 min	plenary
2	Pantomime game	15 min	groups of 4
3	Haptic encoding	20 min	groups of 4
4	Code verification	10 min	groups of 2/individual
	Break	*15 min*	*free*
5	Word de- and encoding	15 min	individual
6	Bitomime	25 min	groups of 4
7	Outro	10 min	plenary

Target Group. The workshop targets primary school students aged 7–10 and requires basic reading and writing literacy, as students work with written symbols, names, and code tables. In particular, students should be able to read, write, and spell their own name, as well as other short words, such as animal names, to fully engage in the encoding and decoding tasks.

Learning Goals. The workshop was designed to introduce students to fundamental principles of information encoding and decoding, variable-length codes, code verification, and binary representation in CS through playful, unplugged learning activities. After successfully completing the workshop, students should be able to:

- Understand that information can be represented through abstract, agreed-upon codes.
- Encode and decode information using instructions or code tables.
- Explain why encoding and decoding are essential processes in CS.

Beyond these conceptual learning goals, the workshop aims to foster broader competencies, including problem-solving, collaborative learning practices, and awareness of how symbolic codes are used in everyday communication. Overall, the workshop seeks to strengthen primary school students' interest in CS and their confidence in engaging with foundational computing concepts.

Workshop Setting. The workshop is conducted as part of the TU Wien informatics outreach program and typically takes place with visiting primary school classes. Students work in groups of four to five and alternate between individual, pair-based, and small-group tasks depending on the activity. The workshop leads (facilitators) guide the activities, provide support when needed, and encourage students to reflect on their learning experiences while connecting the activities to underlying CS concepts through verbal instructions. Facilitation is provided by multiple workshop leads, with an ideal facilitator-to-student ratio of 1:4. The ratio was chosen to ensure continuous individual support during the learning activities and has proven to be effective in prior primary school workshops with similar conditions. However, the workshop can potentially be conducted with fewer facilitators.

3.1 Workshop Activities and Materials

Where letters are encoded, our workshop activities utilize a simplified 5-bit ASCII encoding scheme to encode the Latin alphabet (A to Z): by omitting the first two bits of the regular 7-bit ASCII standard, we end up with a 5-bit scheme that preserves the core logic of binary encoding while reducing complexity. To support a gradual transition to abstract binary notation, binary states are first introduced through tactile 3D-printed pegboards and sticks: a stick can be pinned to a slot on a pegboard, or a slot can be left empty (see Fig. 1). These two states are gradually mapped to the classic states '1' and '0', by first introducing an image-based representation of the pegboard with sticks of only one color (see Fig. 3) and then overlaying the symbolic sticks and slots with digits '1' and '0' (see Fig. 4). The print files for all workshop materials are freely available under a Creative Commons license here: https://doi.org/10.17605/OSF.IO/XH2FE.

In the following we present each step of the workshop.

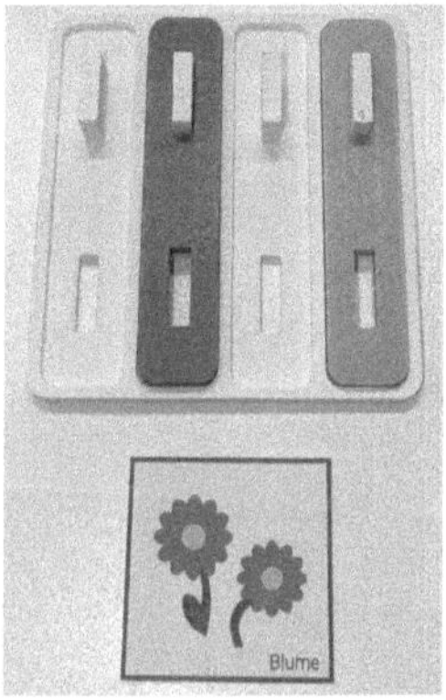

Fig. 1. Pegboard and stick code

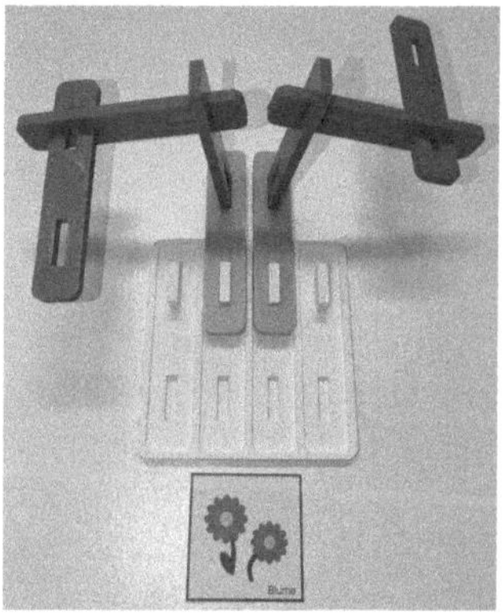

Fig. 2. Iconic pegboards and sticks code

Introduction. The workshop starts with a short, verbal, interactive introduction that situates encoding and decoding as the process of assigning meaning to everyday symbols, emphasized through examples, such as knowing when to cross a street based on the colors of a traffic light. To hook students, multiple rows of 5-bit ASCII codes are displayed on the blackboard. Students are invited to speculate about their meaning, creating curiosity about binary codes as the central topic of the workshop. Through these examples, students are introduced to encoding as the assignment of meaning to signs and symbols, establishing the conceptual foundation for the following workshop activities through contextual examples.

Pantomime game. As the first group activity, students play a charades-like pantomime game. Each student receives a card showing a word or short phrase (e.g. "read a book aloud") or an image including a word describing the image (e.g. a flower, see Fig. 1 for an example). One by one, the students represent the information on their card using only gestures, while the others try to guess its meaning. Through this playful activity, students experience that information can be represented through alternative symbolic systems and the importance of shared understanding in codes, ultimately experiencing the challenges of ambiguity and interpretability, through embodied actions.

Haptic encoding. In the haptic encoding activity, students transfer ideas from the pantomime game into a physical representation using 3D-printed pegboards and sticks in various colors (see Fig. 1). Similar to the previous activity, each student receives a card with a word or image and gets tasked with creating a code to represent it. This time, they create their codes by arranging multiple sticks on a pegboard. In a first open round, students freely construct their representations, often creating three-dimensional, iconic structures (see Fig. 2). When first attempting this, students often report that they are unable to come up with a code for the information on their card using *just* sticks and a pegboard. Workshop facilitators encourage students to do their best, emphasizing that there is no right or wrong. Once every student has come up with a code, each student is asked to explain to the group how their sticks on the pegboard represent the word or image on their card, in other words, how they encoded their information. Students often explain their code in an iconic way, such as "this is the stem, and this is the flower bud", or "the blue stick is water, the yellow on top is a person" when representing *swimming*.

In a second round, the activity is repeated with stricter rules: students may only use up to four sticks and only place each in one of the predefined slots on the pegboard. For students who struggle with this increase in abstraction, workshop facilitators can motivate the new rule by emphasizing its benefits, such as the need for fewer sticks, the better portability of a pegboard were all sticks are secured on the board's pegs, or the ease of reproducing a code by sticking to the slots compared to a more elaborate construction. Additionally, facilitators encourage students to move away from an iconic representation altogether by emphasizing that codes can be chosen arbitrarily: students can simply *decide* that a specific stick arrangement on a pegboard has a particular meaning – no iconicity needed. Figure 1 is an example of a code that adheres to the stricter rules, but has some iconicity left (its colors match the flowers'). Such a result is perfectly acceptable and is built upon during the next activity.

By comparing both rounds, students experience how constraints shape encoding standards and how representations can be simplified and standardized. The binary nature of the pegboard – a slot either contains a stick or not – serves as a visualized bridge toward binary encoding in the following activities and is purposefully emphasized in the workshop leads' verbal instructions.

Code Verification. In this activity, students are introduced to decoding and code verification by analyzing predefined code sequences using a given code table. Students are handed the code verification *worksheet* (see Fig. 3) and *code table* (see Fig. 5) which features pictures of five different cards, similar to the cards used in the pantomime game. Next to each card is its code: a graphical representations of a pegboard-stick code, similar to those used in the haptic encoding activity, but all sticks are of the same color now, ensuring the code is binary. However, to make code verification more challenging, the codes are of varying lengths (2–3 bits), illustrating that encoding schemes do not necessarily rely on fixed-length representations. Additionally, this serves as a bridge from the 4-bit pegboard-stick codes to the later-used 5-bit ASCII codes. Students are tasked to compare several card sequences with their corresponding code sequences and determine whether errors occurred during encoding, using the code table. While two examples are correct, one sequence intentionally contains an error: one bit is missing. The activity highlights the importance of being precise when encoding information using abstract codes: one mistake might already make a code unreadable.

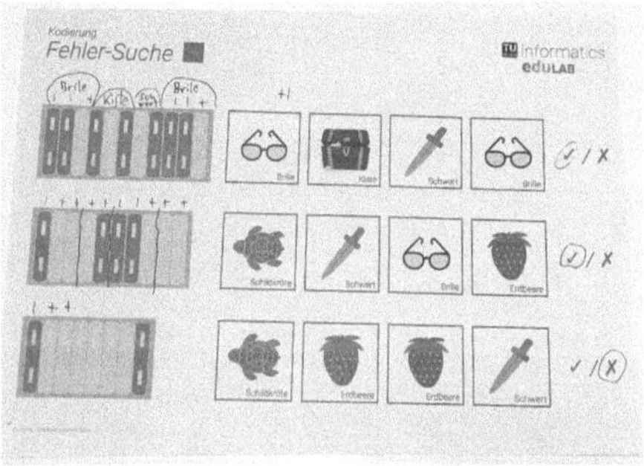

Fig. 3. Solved code verification worksheet

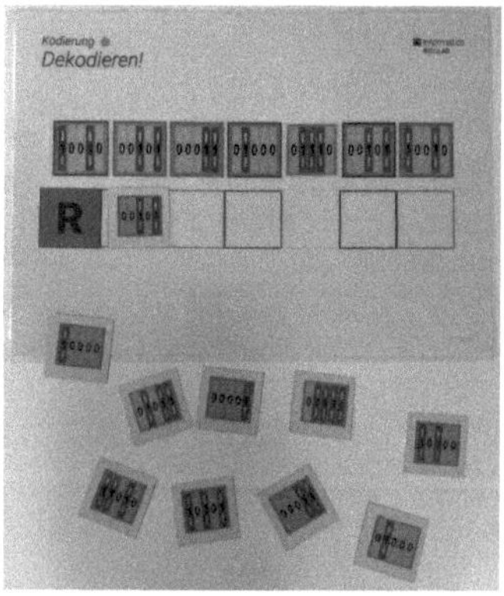

Fig. 4. Decoding worksheet

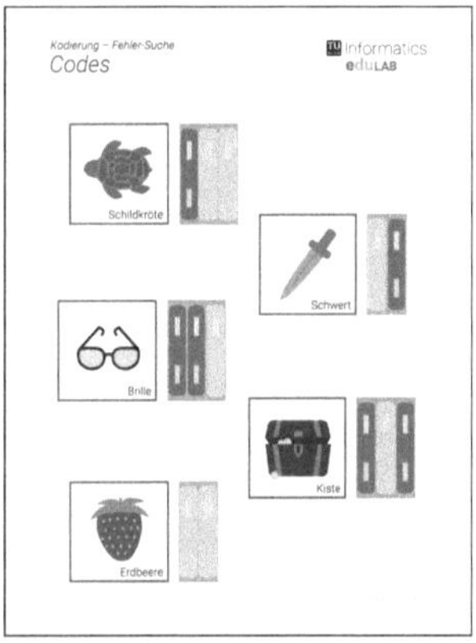

Fig. 5. Code verification code table

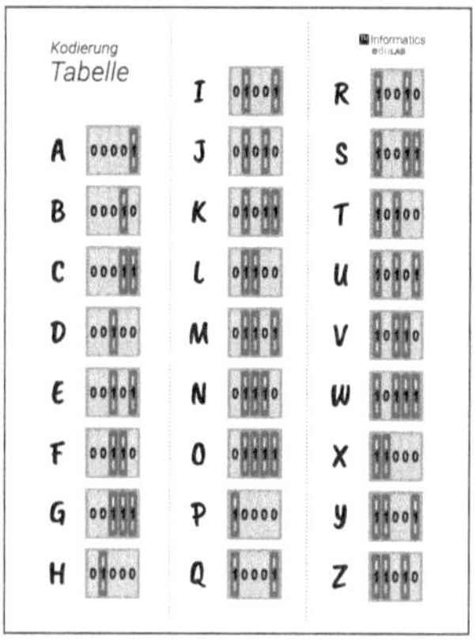

Fig. 6. 5-bit ASCII table

Word De- and Encoding. For the first half of this step, students decode a new binary code sequence. Working in pairs, they receive a worksheet showing seven 5-bit pegboards with sticks and a set of letter cards. This time, the pictures of the pegboards and sticks have the digits '1' printed on each stick, and '0' on each empty slot, to support the transition towards a more abstract binary code. The task, however, can still be solved without engaging with this new representation.

Each letter card displays a 5-bit pegboard code (including ones and zeros) on one side and a letter on the other (see Fig. 4). By matching the cards to the correct pegboards on the worksheet, students gradually decode a seven-letter word and learn that each 5-bit code stands for one letter of the alphabet.

Building on this, students then move from decoding a binary sequence to encoding their own name. Using a 5-bit ASCII table (see Fig. 6), each student

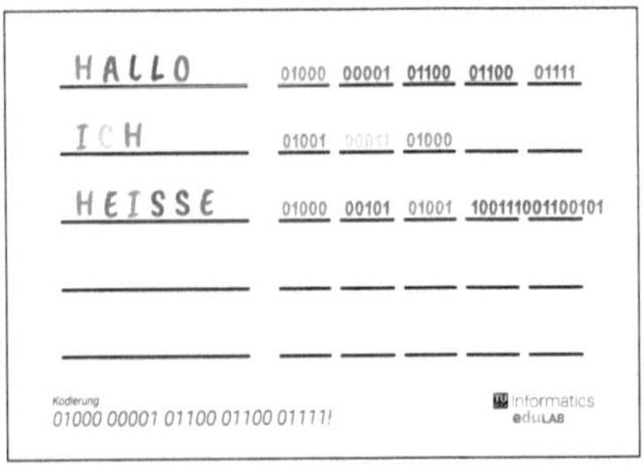

Fig. 7. Name encoding worksheet

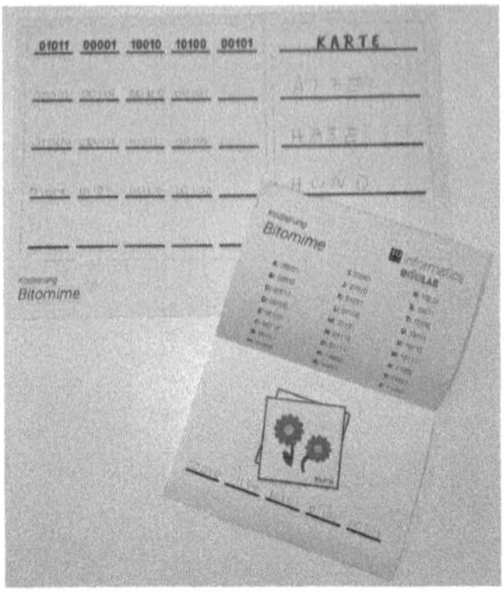

Fig. 8. Filled out Bitomime worksheets

encodes each letter of their name. But rather than using a pegboard or card, students now write zeros and ones on a piece of paper (see Fig. 7). This name encoding worksheet scaffolds students by pre-encoding the phrase "HELLO MY NAME IS", nudging students to finish the phrase using letters on the left side and binary on the right (Fig. 8).

Bitomime. As the final workshop activity, students revisit the initial pantomime game, this time combining it with binary encoding. In *Bitomime*, each student receives a card with a word just like in the pantomime game. To limit the volume of data that has to be transmitted later, only cards featuring words with up to five letters are prepared for this activity. Then, instead of using a gesture to represent their card, students encode the word on their card using the 5-bit ASCII scheme introduced in the previous activity.

The group then agrees on two gestures: one for '0' and one for '1'. Students take turns performing their binary code as a sequence of these two gestures, while the other students write down and decode the sequence to reconstruct the original word. Unlike the initial pantomime activity, the transmission of information now takes place through enacted binary 5-bit ASCII codes. This activity serves as the culmination of the workshop's progression toward abstraction, combining encoding, transmission, and decoding into a single complete digital communication process. By returning to the familiar pantomime format while introducing binary as an intermediate code, students experience how digital communication relies on shared encoding standards to transmit information without ambiguity.

Outro. The workshop concludes with a joint reflection in which students revisit and connect the different en- and decoding activities completed throughout the session. Guided by the workshop lead, students reconstruct the progression from iconic, physical representation in pantomime, through tactile and binary codes, to the complete cycle of encoding, transmitting, and decoding information in

Bitomime, connecting it to the broader concept of encoding in CS. Finally, the binary message displayed on the blackboard is collectively decoded, allowing students to apply and consolidate their newly acquired knowledge, hopefully leaving them empowered to understand encoding as a core concept of CS.

3.2 Experience from Practice

Over the course of a year, the workshop was delivered in 20 individual sessions as part of the TU Wien informatics' outreach program *eduLAB*, reaching a total of 422 primary school students (215 female, 207 male), aged 7–10. Additionally, the workshop was piloted with older students aged 10–14. The workshops were conducted at TU Wien facilities and after each session, the facilitating team conducted oral debriefings to discuss recurring student difficulties, common misconceptions, and the effectiveness of the scaffolding materials. This iterative process led to several refinements of the workshop design.

Student feedback was collected through a traffic-light sticker evaluation at the end of each workshop. Of the 422 students, 335 selected green, 54 yellow, and 8 red, indicating an overall positive reception. This resembles the informal feedback from teachers at the end of the workshops who praised the highly interactive and age-appropriate workshop design, as well as the findings from the workshop leads' informal observations. Overall, the workshop format motivated primary students to engage in encoding and decoding practices throughout the full 120-minute duration. In particular, the progressive transition from embodied activities, starting with a charades-like pantomime game to abstract binary codes seemed to help students gradually build a conceptual understanding of en- and decoding. The final activity, *Bitomime*, repeatedly demonstrated that students successfully combined encoding, transmitting, and decoding information consistent with the workshop's central learning goals.

At the same time, repeated implementation revealed recurring conceptual challenges, which are based on the instructors' informal observations during the workshops and were synthesized through team discussions in relation to the iterative workshop design. While most students were able to successfully encode and decode information by the end of the workshop, it was observable that understanding binary code as an abstract, arbitrary symbolic system remained challenging, particularly for younger students. In the introductory activity, some students correctly associated the code with computers or referred to it as the "computer's language", however, their explanations often remained grounded in familiar decimal number concepts. For some students, notably younger students, the tendency to find iconic or semantic meaning persisted throughout the workshop, as students attempted to assign familiar semantic meaning even to the most abstract versions of codes. This became evident in the semantic connotations of the codes, which they often described, e.g., as "zero, hundred, ten" rather than as a sequence of zeros and ones or e.g., a student stating that the 5-bit code for the letter 'E' looks like an 'E' when turned. A similar pattern emerged in the haptic encoding activity, in which students frequently assigned iconic meaning to the physical arrangement of sticks (e.g., interpreting two vertical sticks as a dog's ears), even when the stricter rules were introduced.

Comparisons across age groups further showed that older students (9–10 years) generally adapted more quickly to the workshop's increasing abstraction and required less scaffolding during encoding and decoding tasks. Younger students (7–8 years), by contrast, benefited more from the embodied and tactile phases and often required additional support during the transition to binary notation, e.g., through haptic pegboards as an explanatory tool during the more abstract encoding activities. While students were able to successfully perform the mechanics of encoding and decoding, the transition from meaningful or iconic representations to arbitrary symbolic codes appears to be a significant conceptual step that requires explicit pedagogical support and reflection. This highlights the didactical relevance of a workshop design that gradually increases abstraction in early encoding/decoding activities and includes opportunities for students to construct their own interpretations of code through collaborative reflections with peers and workshop leads.

Further, the code verification activity revealed recurring difficulties. In particular, students often struggled with code sequences of differing lengths, as this required them to identify boundaries between individual codes. While this complexity supported the intended learning goal of precision in decoding, it also increased the difficulty of the activity, especially for younger students. To scaffold this challenge, workshop leads suggest drawing lines at the end of each code to support mental recognition of individual code sections (see Fig. 3). Similarly, students had challenges receiving the codes in the Bitomime activity. Students often start acting out their gestures very quickly or imprecisely, making it difficult for their peers to interpret and mark down the corresponding ones and zeros. This highlighted the importance of workshop lead guidance in slowing the interaction, emphasizing clarity and precision in symbolic communication, and continuously reiterating encoding as a precise practice. Hence, while the workshop activities can be run with fewer workshop leads, we suggest maintaining tight workshop lead guidance, with a primary focus on the overall organization and timing of the activities, while supporting students' own construction of their encoding and decoding processes.

Further, the workshop design emphasizes the idea of unplugged CS by making intangible abstract concepts tangible through embodiment [14]. The embodied and tactile learning activities can serve as a reference point, highlighting the importance of precision in encoding, as these activities make common challenges visible through students' physical actions.

Based on these repeated implementations, several design principles emerged as central to the workshop's success: (1) gradually increasing abstraction rather than introducing binary notation directly, (2) actively guiding students through the increasing abstraction, emphasizing the arbitrary yet useful notion of binary codes, and (3) introducing abstract concepts through embodied and familiar interaction formats, especially for younger students.

4 Conclusion

This paper presented an unplugged workshop design for introducing primary school students to binary en- and decoding through a sequence of embodied, tactile, and increasingly abstract learning activities. By combining familiar interaction formats, such as pantomime-based games, with tactile codes, code verification, and binary encoding and decoding, the workshop provides an entry point to foundational CS concepts without relying on digital devices.

The repeated implementation of the workshop with over 400 students (aged 7–10), demonstrated that the format was engaging and well-received by students. The workshop leads' observations suggest that the progressive transition from embodied and iconic representations toward abstract binary codes supports students in developing the skill to encode and decode information through encoding instructions or tables, verify codes, and understand that codes are abstract representations of agreed-upon meaning, which is an important practice in CS. At the same time, the workshop revealed that the transition from meaningful or iconic representations toward arbitrary symbolic codes remains a substantial conceptual challenge, especially for younger students. This highlights the importance of gradually moving towards abstraction, repeated reflection, and facilitator support in early CS education.

Future work will explore adaptations for different age groups, such as students in lower secondary education. In particular, more systematic evaluation methods could provide deeper insight into how students develop conceptual understanding of abstract representations and binary encoding over time.

Acknowledgments. This work was partially funded by LEA Project 2WP-0073 on 'Abenteuer Informatik' and WWTF support for female/diversity promotion at TU Wien. All photos and graphics are the work of the TU Wien Informatics eduLAB team. Special thanks to Florentina Gara for additional graphics.

References

1. Bell, T., Lodi, M.: Constructing computational thinking without using computers. Constructivist Found. **14**(3), 342–351 (2019). https://constructivist.info/14/3/342.bell
2. Bell, T., Vahrenhold, J.: CS Unplugged—How Is It Used, and Does It Work? In: Böckenhauer, H.-J., Komm, D., Unger, W. (eds.) Adventures Between Lower Bounds and Higher Altitudes. LNCS, vol. 11011, pp. 497–521. Springer, Cham (2018). https://doi.org/10.1007/978-3-319-98355-4_29
3. Bell, T., Witten, I.H., Fellows, M.: Computer Science Unplugged. Off-Line Activities and Games for All Ages, Lulu (1998)
4. Bers, M.U., Strawhacker, A., Sullivan, A.: The state of the field of computational thinking in early childhood education. OECD Educ. Working Papers 274, OECD (2022). https://doi.org/10.1787/3354387a-en
5. Bundeskanzleramt Österreich: Lehrplan der Volksschule, Anlage 1, Bundesrecht konsolidiert (2025). https://www.ris.bka.gv.at/Dokumente/Bundesnormen/NOR40263891/NOR40263891.html

6. Buttke, H., Krugel, J.: Information Encoding in Computer Science Education Using the Cup Song. In: IEEE Global Engineering Education Conference (EDUCON), pp. 1–10 (2024). https://doi.org/10.1109/EDUCON60312.2024.10578599
7. Computer Science Teachers Association: CSTA K–12 computer science standards, revised 2017 (2017)
8. Grover, S., Pea, R.: Computational thinking in K-12: a review of the state of the field. Educ. Res. **42**(1), 38–43 (2013). https://doi.org/10.3102/0013189X12463051
9. Huang, W., Looi, C.K.: A critical review of literature on "unplugged" pedagogies in K-12 computer science and computational thinking education. Comput. Sci. Educ. **31**(1), 83–111 (2021). https://doi.org/10.1080/08993408.2020.1789411
10. K-12: Computer Science Framework Steering Committee: K-12 computer science framework (2016)
11. Kordaki, M., Mpimpas, C.: Attempts by primary education pupils to learn the concept of binary encoding within the context of drawing: A case study. In: Current Developments in Technology-Assisted Education (2006)
12. Shute, V.J., Sun, C., Asbell-Clarke, J.: Demystifying computational thinking. Educ. Res. Rev. **22**, 142–158 (2017). https://doi.org/10.1016/j.edurev.2017.09.003
13. Vygotsky, L.S.: Mind in Society: The Development of Higher Psychological Processes. Harvard University Press, Cambridge, MA (1978)
14. Waite, J., Maton, K., Curzon, P., Tuttiett, L.: Unplugged computing and semantic waves: analysing crazy characters. In: Proceedings of the 2019 Conference on United Kingdom & Ireland Computing Education Research, pp. 1–7. UKICER '19, Association for Computing Machinery, New York, NY, USA (2019). https://doi.org/10.1145/3351287.3351291
15. Zeng, Y., Yang, W., Bautista, A.: Teaching programming and computational thinking in early childhood education: A case study of content knowledge and pedagogical knowledge. Front. Psychol. **14** (2023). https://doi.org/10.3389/fpsyg.2023.1252718

Teaching AI

From Logic Gates to Hidden Layers: a Hands-on Introduction to Neural Networks for Young Learners

Laura Cesaro[1(✉)], Amy Eguchi[2], Amber Jones[3], Emanuele Menegatti[1], and Thomas Deneux[4]

[1] Department of Information Engineering, University of Padua, Padua, Italy
laura.cesaro@phd.unipd.it , emanuele.menegatti@unipd.it
[2] University of California San Francisco, San Francisco, CA, USA
[3] Georgia Institute of Technology, Atlanta, USA
[4] Institut de Neurosciences Paris-Saclay (CNRS, Université Paris-Saclay), Paris-Saclay, France
thomas.deneux@cnrs.fr

Abstract. Teaching the inner workings of neural networks to primary school children is a powerful way to endow them with an engineering mindset with respect to AI technologies. Yet it requires both appropriate tools and carefully designed educational materials. This paper reports on a pilot experience involving the design, implementation, and evaluation of "Which Food Do You Like?", a hands-on activity in which children aged 10–11 use AlphAI educational software and LEGO Spike robotics to explore the mathematical foundations of supervised learning, from weighted sums and activation thresholds to logic gates and hidden layers, through the lens of a food preference story. The activity was implemented in a primary school context and evaluated through a pre-post questionnaire measuring metacognitive dispositions, AI knowledge, and scenario-based reasoning. Results show both conceptual gains on items directly targeted by the activity and residual misconceptions and difficulties on more complex reasoning tasks. Metacognitive scores improved for all students. We discuss the implications of these findings for the iterative design of the activity and its materials, and make resources openly available.

Keywords: AI Education · Artificial Neural Network · Educational Robotics · Elementary School

1 Introduction

AI literacy is increasingly recognized as a foundational competency of the 21st century, which includes not only the ability to use AI tools but also the understanding of how AI systems are designed, trained on data, and deployed in context [2,6,12–14]. Although neural networks underlie many modern AI systems,

M. Löffler et al. (Eds.): CMSC 2026, LNCS 16951, pp. 117–126, 2027.
https://doi.org/10.1007/978-3-032-38887-2_9

their internal mechanisms remain opaque to non-experts, especially K–12 students. Educational platforms such as Teachable Machine [3], Open Roberta [9] and robotics-based approaches [4,10] have made AI, machine learning, and robotics more accessible through visual and block-based interfaces, but keep the underlying mathematics like weights, thresholds, and intermediate computations, hidden behind user-friendly exteriors. Systems that expose network parameters directly, such as The Neuron Sandbox [5], go further, but do not provide connection to a physical robot or a classroom-ready progression. Educational robotics has also been shown to support students' engagement with abstract computational concepts [7,8]. Making these internal workings tangible for young learners remains a significant educational challenge. The AlphAI platform [1] offers a useful environment to address this challenge by combining a transparent neural network interface with robotic actuation. In this environment, students can observe how sensor inputs are processed by the network, manually edit weights and activation thresholds, train the same network through supervised learning, and see in real time how changes in parameters affect the robot's behavior. Building on this platform, the present paper reports on the design, implementation, and pilot evaluation of an extended hands-on activity aimed at making the mathematical foundations of neural networks visible and meaningful for young learners. Using a food preference story enacted by a physical robot, the activity introduces a cast of fictional characters, each representing a distinct input-output behavior of the network. These characters provide a narrative thread that guides students through weighted sums, activation thresholds, and supervised learning as foundational concepts of neural networks.

2 Methods

2.1 Study Design

This study is strictly exploratory. Rather than testing specific hypotheses about learning outcomes or making generalized claims about effectiveness, we focus on piloting the activity and its associated questionnaire in a real classroom setting. Our specific preliminary aims are to:

- Identify which neural network concepts are most accessible to students.
- Pinpoint where specific difficulties and misconceptions persist.
- Evaluate the effectiveness of the questionnaire as a tool for measuring conceptual change.

The resulting insights will be used to inform future iterative improvements to both the activity design and the evaluation materials.

2.2 Context and Participants

The pilot study involved a class of 11 students aged 10–11 years attending a private primary school in France. Parental informed consent was obtained prior to participation. Complete pre- and post-questionnaire data were available for 9 students; one student completed only the pre-questionnaire and one only the post-questionnaire. The final matched sample therefore consisted of N = 9 students who completed both instruments and were included in all analyses.

2.3 Materials and Activity

Working in pairs, students interact with a LEGO Spike Prime hub equipped with two sensors: a touch sensor to detect a LEGO "pizza" and a color sensor to identify a LEGO "vegetable". The hub displays a happy face when the neural network's output neuron is activated. The AlphAI software runs on a connected laptop and displays the state of the network in real time: input values, weights, pre-activation sums, and activation thresholds, making the intermediate computations visible. Students can interact with the network in two modes: by manually editing weights and thresholds or by collecting labeled examples to train the network through supervised learning. Each of the seven scenarios is introduced through a fictional character whose food preferences define the learning challenge (Table 1). Characters and key concepts are presented through illustrated slides and a storybook narrating the story of a school end-of-year party. The separation between the narrative layer (storybook) and the technical explanation (slides) ensures that conceptual content remains grounded in the robot's behavior rather than in the characters' personalities alone.

Table 1. Seven food-preference scenarios used to introduce neural network concepts, from single weighted inputs to logical gates and hidden-layer representations.

Character	Food preference	Network challenge	Learnt Concept
Luca	Loves pizza	Activate the output when pizza is present	Single input, weight
Isabel	Loves vegetable (pizza does not matter)	Activate the output when vegetables are present	Two inputs, selective connection
Jude	Happy with any food	Activate the output when pizza or vegetables are present	OR gate/combining input
Larry	Is very hungry (needs both to be satisfied)	Activate the output only when both foods are present	AND gate/activation threshold
Maggy	Eats veggy (loves vegetables but cannot eat pizza)	Activate the output for vegetables, but not when pizza is present	Negative weight
Kevin	Is fasting (happy only with no food)	Activate the output only when no food is present	Negative threshold
Alice	Does not mix (one food at a time, never both)	Activate the output when only one food is present	XOR/hidden layer

The seven scenarios listed in Table 1 correspond to Boolean functions of two variables. The objective is not to teach logic gates, but to make the internal functioning of a simple neural network observable: since such functions are few and fully specifiable, each can be verified exhaustively over the four input combinations, so that one network mechanism at a time – weight, activation threshold, negative weight, hidden layer – can be introduced and inspected.

Session 1 —*Tinkering Neural Network Basic Concepts*
In the first phase, students manually set weights and thresholds to reproduce each character's food preference. By manually tuning these parameters, learners construct a transparent, step-by-step understanding of the underlying math and logic of neural networks before moving to automated learning. To provide a proper cognitive scaffolding, the activity begins with a simple single-input challenge (see Fig. 1). This isolates the basic functioning, serving a dual purpose: it ensures students comprehend the fundamental activation process and allows them to familiarize themselves with essential technical terminology (e.g., *input neuron, weight, activation threshold*) without cognitive overload.

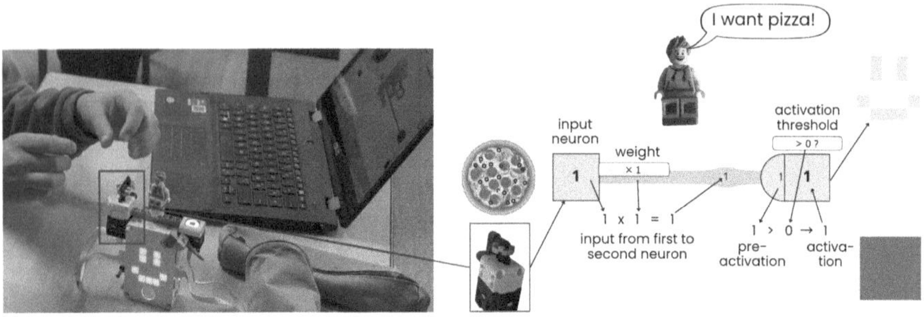

Fig. 1. First contact with an Artificial Neural Network (ANN) with the "Luca wants pizza" challenge. Students manually configure the network by assigning a positive weight to the connection between the *pizza* input neuron and the *smile* output neuron, discovering how to manually trigger the robot's smile according to Luca's preference.

Building on this foundation, students then progress through increasingly complex challenges (working through the preferences of Isabel, Jude, Larry, Maggy, and Kevin). All calculations can be manipulable on-screen at each step. However, managing multiple inputs manually quickly becomes complex and laborious. Students begin to feel the limits of manual editing, as finding the right weights and thresholds requires repeated trial and error.

Session 2—*Supervised Learning and Hidden Layers*
In the second session, students discover a much simpler approach through supervised learning: instead of setting parameters by hand, they switch to learning mode and provide labeled examples. By presenting various food combinations and indicating the desired output, they let the software automatically adjust the weights and thresholds to fit the data (see Fig. 2).

But just as they think they have mastered the system, Alice is introduced. Although she enjoys both types of food, she wants one and only one at a time (an XOR function). Students initially attempted to solve this new challenge using the single-layer network from the previous tasks, but without success. This firsthand experience of failure motivates the introduction of a hidden layer (see Fig. 3),

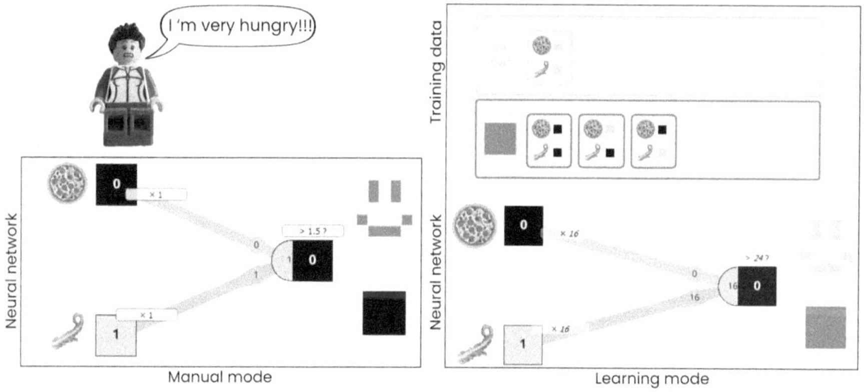

Fig. 2. Revisiting "Larry is very hungry" with Supervised Learning. Left: Solving "Larry is very hungry" by manually editing connections and activation bias (setting weights of 1 and a threshold > 1.5 solves the challenge, since it will ensure activation if and only if both inputs are active). Right: Solving it through Supervised Learning. Students record for which sensor states the robot should smile (when both sensors are activated) or not smile (when no sensor or only one sensor is activated), and the AI auto-adjusts weights to find a solution that "works", i.e., that will activate the smile exactly on the specified sensor state(s).

showing students how more complex networks can solve problems that single-layer architectures cannot–giving them an intuition on how real-world neural networks work, with many more layers and neurons, while reinforcing that their outputs, however complex, ultimately stem from the fundamental arithmetic operations they have been performing by hand, replicated across millions of parameters.

Both sessions were highly participative. Students edited weights and thresholds in the AlphAI software and immediately tested the resulting behavior on the LEGO robot. After each activity step, one or two students worked through the network's computations at the whiteboard – multiplying each input by its weight, summing the products, and comparing the result with the activation threshold—in all 4 cases of presence/absence of pizza/vegetable. These moments served as a whole-class recap, in which the procedure was retraced for all students and not only for those at the board, allowing the class to reason about the logic underlying the software interface.

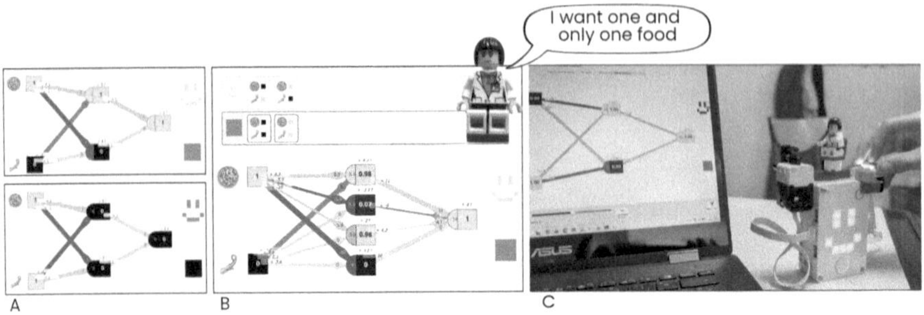

Fig. 3. Introducing a hidden layer to solve Alice's "one and only one food" XOR challenge. (A) After unsuccessful attempts with the single-layer network, students manually configure a hidden layer with two intermediate neurons. They follow a modular logic: the first hidden neuron activates only for "pizza but no vegetables", the second only for "vegetables but no pizza", while the output neuron performs an OR gate. (B) Students solve the challenge again using Supervised Learning, observing that the algorithm converges more effectively when adding more intermediate neurons (four as shown here). (C) The physical classroom setup, demonstrating the real-time synchronization between the software interface and the LEGO robot hub's response.

2.4 Data Collection

Students completed the questionnaire before and after the activity. The questionnaire included metacognitive Likert-type items and AI knowledge-based items. The metacognitive items were included because the activity invited students to reflect on learning processes—such as recognizing errors, using feedback, and adjusting one's understanding—that were conceptually related to the idea of learning from data during AI training. Higher scores indicated greater agreement with the metacognitive statements and greater accuracy in the knowledge-based items. The questionnaire was organized into thematic sections, combining items adapted from Martin et al. [11] with items developed specifically for this activity. Table 2 summarizes the structure of the instrument, the content of each section, and whether the items were administered before and/or after the intervention.

The pre- and post-questionnaires were identical across the shared items (Q1–Q17), allowing direct comparison before and after the intervention. The post-

Table 2. Structure of the questionnaire items, including their thematic focus, origin, and administration phase.

Items	Section/focus	Origin	Administration
Q1–5	Metacognitive knowledge	Adapted from Martin et al.	Pre and post
Q6–13	General knowledge about artificial intelligence	Adapted from Martin et al.	Pre and post
Q14–17	Logic of AI and neural networks	Developed for this activity	Pre and post
Q18–22	Post-activity application items	Developed for this activity	Post only
Q23–25	Ethics and limits of AI decision-making	Developed for this activity	Post only

only items (Q18–Q25) were used to assess students' ability to apply the concepts introduced during the activity and to reason about ethics and the limits of AI decision-making.

3 Results

As outlined in the objectives of the pilot study, we first present the descriptive analyses and pre-post trends to gain initial insights into the intervention's potential. Two composite scores were computed by averaging item scores: a metacognitive learning score based on items 1–5 and an AI knowledge score based on items 6–13. Item 17 was excluded from the analysis due to ambiguous wording that may have led to inconsistent interpretation across students. Descriptive statistics were computed for both composite scores at pre-test and post-test. Given the small sample size and the paired pre-post design, changes in the two composite scores were examined using two-sided Wilcoxon signed-rank tests. Item-level results were inspected descriptively only, to inform future refinement of the questionnaire.

Metacognitive Beliefs (Q1–Q5). Students' metacognitive scores improved from pre- to post-test (M pre = 0.82; M post = 0.92; $p = .002$).

AI Knowledge (Q6–Q13). AI knowledge scores also improved from pre- to post-test (M pre = 0.73; M post = 0.85; $p = .004$). The largest significant gains were observed on Q10 ("In order to learn, an artificial intelligence must be trained", p = .008). A positive trend was also observed for Q8 ("Machines that operate with artificial intelligence are programmed", p = .063), though it did not reach statistical significance.

Logic of AI and Neural Networks (Q14–Q16). Q14 asked students to identify who is responsible when an AI makes a mistake (multiple answers possible: the AI programmer, AI itself, the AI user). At pre-test, 3 out of 9 students selected all three options; at post-test, 1 student selected all three, with most students attributing responsibility primarily to the programmer. Q15 (whether a pizza sensor would influence a robot's decision for a character who loves pizza) improved from M = 0.50 to M = 0.83.Âă Q16 (identifying the robot's correct output when both pizza and vegetables are present for Maggy) showed no change (pre = post = 0.44).

Post-only Items (Q18–Q25). Post-only items assessed students' ability to apply the concepts introduced during the activity across three areas: reading neural network diagrams, reasoning about training data, and reflecting on the ethical limits of AI decision-making.

- **Diagram reading (Q18–Q20b).** Q18 asked whether Alice would be happy given a diagram of her XOR network configuration (M = 0.67). Students were then shown two further diagrams and asked to compute the robot's output and identify the corresponding character. For the Luca diagram (pizza weight = 1, vegetable weight = 0), 78% correctly determined that the robot would

not smile when only a vegetable is present, and 67% correctly identified the character as Luca. For the Kevin diagram (both weights = –1, threshold > -1), 44% correctly determined that the robot smiles when no food is present, although 89% correctly identified the character as Kevin.
- **Training data reasoning (Q21–Q22).** Students were asked whether a given set of training examples would be sufficient to capture Maggy's food preferences (Q21: M = 0.78) and whether a given dataset would be appropriate for training Alice's XOR behavior (Q22: M = 0.89).
- **Ethics (Q23–Q25).** Students evaluated three statements about the limits of the robot's knowledge and decision-making. Performance was high on Q23 (M = 0.89) and Q25 (M = 1.00), but lower on Q24 (M = 0.56), asking whether the robot knows that it is not good for Kevin to eat nothing.

Qualitative Observations of Students. After each of the 7 activity steps one or two students were invited to the whiteboard to retrace all the mathematical details of operations inside the neural network in all 4 cases of presence/absence of pizza/vegetable. All of the 9 students were able to speak and explain to the class. 6 of them provided the correct explanation without help from the teacher, whereas 3 appeared more unsure and could do it only with the teacher's support. Altogether, the advantage of using a concrete Lego setup with robot and characters was clearly palpable: kids were enthusiasts, eager to participate, including presenting to the rest of the class at the whiteboard; and teachers could witness that several mathematical notions that they first presented at the whiteboard were clearly understood by the students only once they were manipulating them in the software (changing parameter values) and physically (placing/removing pizza and vegetable on the robot's sensors and in the software interface and observing the robot face).

4 Discussion

The pilot results provide preliminary support for the proposed activity as a way to make selected mathematical ideas underlying neural networks accessible to young learners. Both composite scores showed statistically significant pre-post improvements, despite the small sample size. At the level of individual questions, the results offer useful insights into how the activity may be refined. Responses to Q14 showed a shift at post-test, with fewer students selecting all three responsible parties and more students attributing responsibility primarily to the programmer, proving their good new understanding of the dependence of AI on training, which was one of the key teaching in this activity. Of course the broader issue of shared responsibility in AI systems (including e.g. user responsibility in mis-using the system) may need to be addressed more explicitly in future iterations.

The scenario-based questions were particularly informative for evaluating whether students could apply the mathematical ideas introduced through the narrative. Q18–Q19 showed that they mastered all the needed concepts: weight multiplication, summation of multiple inputs, threshold comparisons, multi-layer serialization. However, the manipulation of negative numbers appeared to pose some difficulties to students, as indicated by weaker scores to Q20, where they struggled more with computing the diagram output than with identifying the character represented. This is in accordance with the absence of pre-post change in Q16, suggesting that negative weights remained difficult for students. Note that we also found out that one question in our forms (Q17) needed to be reformulated as its wording was not clear enough.

Altogether our results show that the narrative and character-based structure supported students' interpretation of the network behaviors, while the underlying calculations, especially those involving negative weights which are naturally challenging at this age, may require additional scaffolding. These preliminary findings must be interpreted cautiously due to the small scale of the pilot, yet were confirmed by qualitative obserations of the class. Altogether, this pilot study provided clear guidance for refining the assessment tools and confirmed the value of the instructional design.

In addition to the proved acquisition of new knowledge and skills in AI, students displayed improvements in their metacognitive attitudes toward learning, indicating the benefit of practical manipulations. This also will need confirmation by experimenting on a larger students group.

5 Conclusion

This pilot study suggests that the proposed hands-on, narrative-based activity can support young learners' engagement with foundational ideas of neural networks, including weighted sums, activation thresholds, and AI training. The pre-post changes observed in the composite scores are promising, while the identification of areas requiring further scaffolding, such as calculations involving negative weights, and questionnaire items needing refinement provides clear guidance for future implementations and larger-scale evaluations. By making the underlying mathematics tangible rather than hiding it behind a black box, this approach lays a solid foundation for developing robust AI literacy in primary education.

Generative AI Disclosure. The authors used an AI-assisted editing tool to improve grammar, clarity, and readability during manuscript preparation.

References

1. Absalon, M., Deneux, T.: AlphAI: Teaching ai algorithms to k-12 by training learning robots and visualizing how it works, pp. 29096–29103. AAAI (2025)
2. AI4K12 Initiative: Ai4k12: Sparking curiosity in ai. AAAI and CSTA (nd). https://ai4k12.org
3. Carney, M., et al.: Teachable machine: Approachable web-based tool for exploring machine learning classification. In: Extended Abstracts of the 2020 CHI Conference on Human Factors in Computing Systems, pp. 1–8. ACM, New York, NY, USA (2020)

4. Cesaro, L., Dodero, G., Menegatti, E.: Integrating ai into school curriculum: A maker-oriented activity. In: Balogh, R., Obdržálek, D., Fislake, M. (eds.) Robotics in Education, pp. 367–378. Springer Nature Switzerland, Cham (2024)
5. Chen, A., Pawar, N., Touretzky, D.: Neuron sandbox (2024). https://www.cs.cmu.edu/~dst/NeuronSandbox/
6. Cosgrove, J., Cachia, R.: Digcomp 3.0: European digital competence framework. Tech. rep., Publications Office of the European Union, Luxembourg (2025). https://publications.jrc.ec.europa.eu/repository/handle/JRC144121
7. Eguchi, A.: AI-Powered Educational Robotics as a Learning Tool to Promote Artificial Intelligence and Computer Science Education. In: Merdan, M., Lepuschitz, W., Koppensteiner, G., Balogh, R., Obdržálek, D. (eds.) RiE 2021. AISC, vol. 1359, pp. 279–287. Springer, Cham (2022). https://doi.org/10.1007/978-3-030-82544-7_26
8. Eguchi, A.: Revisiting the pedagogy of educational robotics. In: Balogh, R., Obdržálek, D., Christoforou, E. (eds.) International Conference on Robotics in Education (RiE), Lecture Notes in Networks and Systems, Vol. 747, pp. 81–92. Springer, Cham (2024). https://doi.org/10.1007/978-3-031-38454-7_8
9. Jost, B., Ketterl, M., Budde, R., Leimbach, T.: Graphical programming environments for educational robots: Open roberta-yet another one? In: Proceedings of the IEEE International Symposium on Multimedia, pp. 381–386. IEEE (2014)
10. Karalekas, G., Vologiannidis, S., Kalomiros, J.: Teaching machine learning in k-12 using robotics. Educ. Sci. **13**(1), 67 (2023)
11. Martin, M., Chevalier, M., Burton, S., Bonvin, G., Besançon, M., Deneux, T.: Effects of introducing a learning robot on the metacognitive knowledge of students aged 8–11. In: Balogh, R., Obdržálek, D., Christoforou, E. (eds.) International Conference on Robotics in Education (RiE), pp. 169–183. Springer, Cham (2023). https://doi.org/10.1007/978-3-031-38454-7_15
12. OECD: Education at a Glance 2023: Oecd indicators. Tech. rep., OECD Publishing, Paris (2023). https://doi.org/10.17871/e13bef63-en. https://www.oecd.org/en/publications/education-at-a-glance-2023_e13bef63-en.html
13. OECD: Empowering learners for the age of ai: An ai literacy framework for primary and secondary education. Review draft, OECD, Paris (2025). https://ailiteracyframework.org
14. UNESCO: AI competency framework for students. UNESCO, Paris (2024). https://www.unesco.org/en/articles/ai-competency-framework-students

The Fairness Puzzle: Unplugged Activities for Discovering Why Fair AI is Hard

Antonio Rodà(✉)

Department of Information Engineering, University of Padova, Via Giovanni Gradenigo 6/B, 35131 Padova, Italy
antonio.roda@unipd.it

Abstract. The rapid adoption of Artificial Intelligence has brought increasing attention to the concepts of fairness and non-discrimination. In the European Union, these concerns are reflected both in the Ethics Guidelines for Trustworthy AI, which identify diversity, non-discrimination and fairness as one of seven key requirements, and in the EU AI Act, which establishes risk-management and data-governance obligations for high-risk AI systems. In this contribution, I present two unplugged activities developed for the undergraduate course Gender Knowledge and Ethics in Artificial Intelligence, a general course of the University of Padova. The course is attended by students from diverse disciplinary backgrounds and therefore requires approaches that avoid excessive formalism while preserving conceptual rigor. The first activity uses coloured cards, short applicant profiles, and scholarship tokens. Participants apply a simple, explicitly provided scoring rule and then audit its outcomes through counting and comparison. The second activity is structured as a puzzle inspired by Sudoku. Participants are asked to complete a grid under a set of fairness constraints that appear reasonable when considered individually. Through collaborative problem solving, they eventually realize that the puzzle has no solution. This discovery provides an intuitive introduction to impossibility results in algorithmic fairness and highlights the inherent trade-offs between competing notions of fairness. Both activities follow the Computer Science unplugged approach and require no programming background.

Keywords: Fair AI · Responsible AI · Unplugged activities · AI education · Fairness · Fairness metrics

1 Introduction

Artificial Intelligence (AI) systems increasingly influence decisions that affect people's lives. Machine learning models are routinely employed in domains such as social networks, entertainment, news, recruitment, healthcare, education. While these systems offer opportunities for efficiency and scalability, they also raise important concerns regarding fairness, discrimination, and accountability.

M. Löffler et al. (Eds.): CMSC 2026, LNCS 16951, pp. 127–138, 2027.
https://doi.org/10.1007/978-3-032-38887-2_10

Numerous studies have documented cases in which algorithmic systems reproduced or amplified existing social inequalities, producing systematically different outcomes for individuals belonging to different demographic groups [2,4,10].

The growing awareness of these issues has led fairness to become a central topic in both AI research and public policy. In the European Union, diversity, non-discrimination and fairness are explicitly identified as one of the seven key requirements for Trustworthy AI proposed by the High-Level Expert Group on Artificial Intelligence [7]. More recently, the EU AI Act has introduced risk-management and data-governance obligations for high-risk AI systems, including measures intended to identify and mitigate discriminatory outcomes and bias [5]. Consequently, understanding fairness is no longer a concern restricted to computer scientists and machine learning researchers; it is becoming an essential component of digital citizenship and AI literacy.

Despite its importance, fairness remains a challenging concept to communicate. One reason is the absence of a single, universally accepted definition of fairness: different notions of fairness reflect different perspectives on what a fair decision should achieve and may even conflict with one another. For example, consider an AI system used to award scholarships. One view of fairness would require that the same percentage of male and female applicants receive a scholarship. Another view would require that qualified applicants have the same chance of being selected regardless of gender. The two principles focus on different goals: the first seeks equal outcomes across groups, while the second seeks equal chances for qualified individuals. In some situations, satisfying one principle may make it impossible to satisfy the other, illustrating that fairness is not a single concept but a collection of competing criteria. The research literature has proposed a wide variety of fairness definitions, including demographic parity, equal opportunity, equalized odds, predictive parity, individual fairness, and counterfactual fairness [6,9]. Each definition captures a different intuition about what it means for a decision-making system to be fair. Furthermore, a seminal result by Kleinberg et al. [8] demonstrated that several desirable fairness criteria cannot generally be satisfied simultaneously when demographic groups exhibit different *base rates*, meaning that the groups contain different proportions of individuals who should legitimately receive a positive outcome by the system, such as obtaining a loan, being hired, or being admitted to a programme.

This complexity creates significant educational challenges. Discussions of fairness are often framed either as ethical debates, from a philosophical and/or social perspective, disconnected from computational mechanisms or as highly technical topics, from a computer science and/or data science perspective, requiring advanced knowledge of statistics and machine learning. Existing surveys provide extensive taxonomies of fairness definitions and mitigation methods [6,9]. However, comparatively little attention has been devoted to reusable unplugged activities through which learners can manipulate the quantities underlying fairness metrics and discover their incompatibilities directly. As AI systems increasingly affect society, there is a growing need for educational activities that make the mathematical dimensions of fairness accessible to broader audiences. The challenge is particularly relevant in interdisciplinary educational contexts. The activities presented in this paper were developed within the undergraduate course *Gender Knowledge and Ethics in Artificial Intelligence* at the University of Padova [1]. The course is offered as a General Course within an engineering curriculum but attracts students from a wide variety of disciplinary backgrounds, including social sciences, humanities, and law studies. Consequently, traditional approaches based on formal definitions, statistical derivations, or machine learning implementations are not always effective in supporting conceptual understanding.

To address this challenge, an unplugged card-based activity was introduced in the previous edition of the course. Inspired by the Computer Science Unplugged philosophy [3], the activity used colored cards representing different demographic groups and decision outcomes within a mortgage approval scenario. Students were asked to evaluate the fairness of an AI system deciding whether to grant or deny loans, exploring fairness metrics through direct manipulation and counting. Student feedback was highly positive: in the final course evaluation, many participants identified the activity as one of the most useful learning experiences for understanding both the importance of fairness in AI and the difficulties involved in defining and achieving it.

Building on this experience, the activity was redesigned, extended, and formalized into a broader educational framework. During this redesign, the original mortgage scenario was replaced with a scholarship allocation scenario, as it was considered closer to the students' own experiences and therefore more likely to foster engagement and discussion. This paper presents the resulting proposal, which consists of two unplugged activities aimed at introducing central concepts in algorithmic fairness. The first activity further develops the original card-based approach, allowing participants to compare alternative fairness criteria through hands-on exploration. The second activity, called the *Fairness Puzzle*, transforms fairness constraints into a collaborative constraint-satisfaction problem inspired by Sudoku. Through repeated attempts to satisfy multiple fairness requirements simultaneously, participants discover that some combinations of constraints admit no solution. The activities described in this paper have not yet been implemented in their current form. Rather, they should be understood as the design of two learning units that will be introduced and evaluated during the next edition of the course.

The main contribution of this work is not the simplification of fairness concepts but rather the design of learning experiences that allow participants to discover fairness trade-offs for themselves. By transforming abstract fairness metrics into tangible mathematical objects, the proposed activities help learners understand that fairness in AI is both an ethical concern and a problem of mathematical modeling.

2 Activity 1: Fairness with Cards

Learning Objectives. The activity, that is divided in four phases (Table 1), aims to introduce participants to the concept of algorithmic fairness and to the idea that different fairness criteria may lead to different interpretations of what a fair decision should be. More specifically, participants are expected to: (i) recognize that people and institutions may use AI systems to support or automate decisions affecting groups of individuals; (ii) understand the meaning of basic classification outcomes such as correct and incorrect decisions; (iii) compare different notions of fairness; and (iv) reflect on the ethical implications of adopting one fairness criterion rather than another.

Target Audience and Prerequisites. The activity was designed for undergraduate students attending the course *Gender Knowledge and Ethics in Artificial Intelligence.* No prior knowledge of statistics, machine learning, or programming is required. Since the activity focuses on reasoning about decisions, constraints, and trade-offs rather than on mathematical formalism or algorithmic implementation, it can be readily adapted to a wide range of non-specialist audiences. Examples include teacher professional development workshops, corporate training programmes on responsible AI, and outreach activities aimed at promoting computational thinking and AI literacy among citizens.

Materials. The activity uses a deck of cards representing individuals, a set of short applicant-profile slips, a printed scoring rule, and scholarship tokens. The front of each card identifies an abstract demographic group (blue or red); these colours are arbitrary group labels[1]. Each profile slip reports a small number of visible, non-sensitive attributes, such as an academic score, attendance, and an interview score. The printed rule combines these attributes into a recommendation, thereby avoiding random selection while keeping the computation accessible. The back of each card contains the hidden reference outcome: whether the applicant would succeed if awarded the scholarship. Tokens indicate positive recommendations. Figure 1 shows the card and token design and a simplified example of play. The profile slips and scoring-rule sheet are deliberately simple additions to the illustrated materials and can be adapted to the audience.

Expected Learning Outcome. At the end of the activity, participants should be able to explain why different definitions of fairness may produce different outcomes and why fairness in AI cannot always be reduced to a single universally accepted criterion.

Phase 1: Introducing the Scenario. Participants are presented with a simplified decision-making problem. A fictitious university must assign a limited number of scholarships with the support of an AI-based scoring system. Each group receives cards representing applicants, the corresponding profile slips, a printed scoring rule, and scholarship tokens. The front of each card identifies the applicant's abstract demographic group, while the profile provides the attributes available

[1] The colours used to identify the two demographic groups have no semantic meaning. If blue and red carry specific cultural connotations (e.g., blue associated with correctness and red with errors), they can be replaced by any pair of neutral colours.

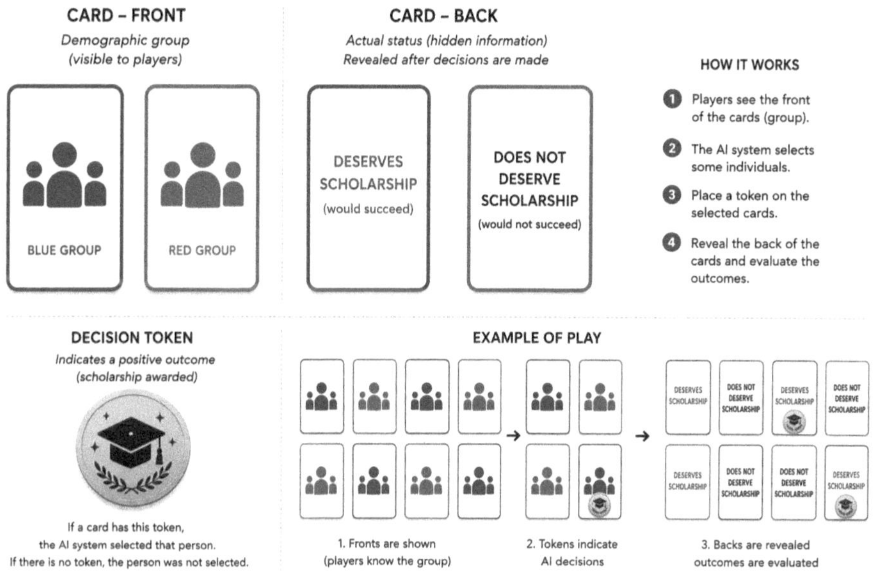

Fig. 1. Example of the materials used in the card-based activity. The front identifies the demographic group, while the back contains hidden information about whether the individual would actually benefit from receiving the scholarship. Scholarship tokens represent the decisions made by the simulated AI system.

to the system. The back contains hidden information indicating whether the student would actually succeed in the programme if awarded a scholarship.

The facilitator explains that the scoring rule is a deliberately transparent stand-in for a trained model: participants do not choose recipients at random or solely from group membership, but apply the same rule to the visible profile of every applicant. The demographic group is retained only so that the resulting decisions can later be audited across groups. The hidden reference outcome is not used when the recommendation is produced.

Phase 2: Simulating Algorithmic Decisions. Working in small groups, participants calculate the score specified by the printed rule and place a scholarship token on every card whose score exceeds the stated threshold. Cards without a token represent applicants not recommended by the simulated system. Because the rule is fixed and applied to visible applicant attributes, different groups should initially obtain the same recommendations; their task is to execute and subsequently audit the rule rather than to imitate an opaque or random AI.

Once all decisions have been made, the cards are turned over and the hidden information is revealed. Participants then compare the AI decisions with the actual outcomes and classify each case as a True Positive (TP), False Positive

Table 1. Copyable activity chart for the card-based activity.

Phase	Organisation	Main task	Suggested time
1	Whole class	Introduce the scholarship scenario, profile attributes, and scoring rule.	10 min
2	Small groups	Apply the rule, place tokens, reveal outcomes, and count TP, FP, TN, and FN.	20 min
3	Small groups	Audit the result and reallocate tokens under alternative fairness criteria.	20 min
4	Whole class	Compare solutions and discuss ethical priorities and trade-offs.	15 min

(FP), True Negative (TN), or False Negative (FN). For example, a student who receives a token and is revealed to deserve the scholarship corresponds to a True Positive, whereas a student who receives a token but would not actually succeed corresponds to a False Positive.

The numbers of TP, FP, TN, and FN are then counted separately for each demographic group. These values provide the basis for discussing different fairness criteria in the following phase. Participants are encouraged to reflect on the quality and fairness of their decisions and to compare the outcomes obtained by different groups.

Phase 3: Exploring Fairness Criteria. The facilitator introduces two fairness criteria.

The first criterion, *Demographic Parity*, requires that the same proportion of students be selected from each demographic group. The second criterion, *Equal Opportunity*, requires that students with the same level of merit have the same probability of being selected regardless of their group membership.

Participants first calculate whether the initial recommendations satisfy each criterion. They then use the revealed qualification labels to conduct a second, explicitly problem-solving round: while keeping the total number of scholarships fixed, they move tokens to construct one allocation satisfying Demographic Parity and another satisfying Equal Opportunity. In many cases, they discover that a solution satisfying one criterion does not satisfy the other. The abstract blue and red labels are used on the physical cards to avoid assigning value to a real social group; during the discussion, the facilitator may explain that an audit in practice could compare legally or socially relevant groups, such as women and men.

Phase 4: Reflection and Discussion. The activity concludes with a guided discussion. Participants analyse the consequences of adopting different fairness criteria and reflect on the values underlying each choice. The discussion highlights that fairness is not only a technical property of an AI system but also a normative concept linked to different social, ethical, and political priorities.

Methodological Considerations. Rather than introducing formal definitions at the beginning, participants encounter fairness as a concrete decision-making problem and progressively construct the relevant concepts through direct experience. Mathematical formalization is introduced only after participants have developed an intuitive understanding of the underlying issues.

Table 2. Comparison of two allocations for the blue and red groups.

Allocation	Group	Selected	Selection rate	Qualified	True Positive Rate
Demographic Parity	Blue	5/20	25%	8	depends on selected cases
	Red	5/20	25%	12	depends on selected cases
Equal Opportunity	Blue	4 qualified	may differ	8	50%
	Red	6 qualified	may differ	12	50%

Table 2 provides two numerical examples designed to help participants understand how different fairness criteria can lead to different decision-making outcomes. Demographic Parity focuses on equal selection rates across groups. Equal Opportunity focuses instead on equal true-positive rates, requiring qualified applicants in the two groups to have the same probability of selection. The table uses the same abstract blue and red group labels as the cards; women and men are introduced only later as a possible real-world interpretation in the numerical puzzle.

3 Activity 2: The Fairness Puzzle

Learning Objectives. The activity aims to help participants discover that different fairness criteria may be mutually incompatible. More specifically, participants are expected to: (i) interpret fairness metrics as constraints on a decision-making system; (ii) understand that different fairness definitions can coexist in tension; (iii) experience a simplified fairness impossibility result through collaborative problem solving; and (iv) recognize that fairness involves both mathematical and ethical choices.

Target Audience and Prerequisites. The activity is intended for participants who have already been introduced to basic fairness concepts, for example through the card-based activity described in the previous section. No advanced mathematical knowledge is required.

Materials. Participants receive a worksheet containing two confusion matrices corresponding to two demographic groups (e.g., women and men), a set of fairness constraints, and small counters labelled TP, FP, TN, and FN. Rather than repeatedly erasing numbers, groups place and move counters in the matrix cells and record the final values only when they believe that all constraints have been met. The task is to complete the tables while satisfying all the constraints simultaneously.

Expected Learning Outcome. At the end of the activity, participants should understand that apparently reasonable fairness requirements may be impossible to satisfy simultaneously and that the choice of a fairness metric often implies trade-offs.

Phase 1: Introducing the Puzzle. Participants are presented with a simplified AI system used to award scholarships. Two demographic groups are considered. For each group, the total number of qualified and non-qualified applicants is known.

An incomplete confusion matrix is provided for each group. Participants are informed that the AI system must satisfy a set of fairness requirements and are asked to complete the missing values.

Table 3. Fairness Puzzle worksheet presented at the start of Activity 2.

Group	TP	FP	TN	FN
Women	?	?	?	?
Men	?	?	?	?

As in Sudoku, the entries are not chosen independently: row totals are fixed by the numbers of qualified and non-qualified applicants, while each fairness condition imposes an additional relation between cells in different rows. A value placed in one cell therefore restricts the admissible values elsewhere. The objective is not merely to fill the table but to determine whether a configuration satisfying every condition exists.

Phase 2: Solving the Puzzle. Working in small groups, participants attempt to construct a solution satisfying all the constraints. Typical constraints include:

- Equal Opportunity: the same proportion of qualified applicants must be selected in both groups;
- Equalized Odds: both true positive rates and false positive rates must be equal across groups;
- Predictive Parity: a positive decision must have the same reliability across groups, meaning that the proportion of selected applicants who are actually qualified must be equal in the two groups.

Participants are encouraged to use trial-and-error strategies and to justify their reasoning. In most cases, several partial solutions emerge, each satisfying some but not all of the constraints.

Phase 3: Discovering the Impossibility. After a period of exploration, participants typically realize that no configuration satisfies all the constraints simultaneously. The facilitator then guides a collective discussion to analyse the failed attempts and identify the source of the contradiction.

The discussion introduces the idea that fairness metrics can be interpreted as competing requirements imposed on the same decision-making process. When demographic groups differ in the proportion of individuals who would genuinely benefit from the decision, some fairness criteria become mathematically incompatible.

Phase 4: Connecting the Puzzle to AI Fairness. The facilitator concludes by connecting the puzzle to contemporary debates on algorithmic fairness. Participants are shown that the difficulty encountered during the activity mirrors a well-known result in the fairness literature, namely that several desirable fairness criteria cannot generally be satisfied simultaneously [8].

Rather than presenting the impossibility result as a theorem, the activity allows participants to experience it directly through exploration and discussion. In this way, the mathematical structure of the problem emerges naturally from the participants' attempts to solve the puzzle.

Methodological Considerations. The Fairness Puzzle adopts the structure of a constraint-satisfaction problem inspired by logic games such as Sudoku. The analogy concerns the propagation of constraints rather than the visual shape of the worksheet: fixed row totals and fairness equations play the role of Sudoku clues, so each placement restricts the values that can be placed in other cells. This format was chosen because it shifts the focus from memorizing fairness definitions to actively exploring their consequences. Participants engage in mathematical reasoning without requiring formal proofs, while the collaborative nature of the activity promotes discussion about the ethical and social implications of different fairness criteria. The puzzle therefore serves as a bridge between computational thinking and ethical reflection, illustrating how mathematical constraints can shape real-world decisions made by AI systems.

Numerical Examples. The activity is framed as the design of an AI system for awarding scholarships. Participants are informed that the applicant pool consists of two demographic groups, women and men, and that the true qualification status of the applicants is known. In the proposed example, there are 20 women applicants, of whom 8 are qualified and 12 are not, and 20 men applicants, of whom 12 are qualified and 8 are not. Participants are then asked to complete the confusion matrices (Table 3) of the AI system by determining the numbers of true positives (TP), false positives (FP), true negatives (TN), and false negatives (FN) for each group.

The goal is to construct a decision-making system that satisfies a set of fairness requirements simultaneously. Participants begin by placing 8 qualification counters and 12 non-qualification counters in the women's row, and 12 qualification counters and 8 non-qualification counters in the men's row; they then distribute each total between the positive and negative decision cells.

The fairness requirements are the following:

- Equal Opportunity: qualified applicants should have the same probability of receiving a scholarship in both groups.
- Equalized Odds: the AI system should make the same proportion of false positive decisions in both groups.
- Predictive Parity: among the applicants receiving a scholarship, the proportion of truly qualified recipients should be the same in both groups.

For the numerical values used in this example, Equal Opportunity requires a true positive rate of 0.5 in both groups, while Equalized Odds requires a false positive rate of 0.25 in both groups. These two constraints uniquely determine all entries of the confusion matrices, leading to the solution shown in Table 4.

Table 4. Unique solution satisfying Equal Opportunity and Equalized Odds.

Group	TP	FP	TN	FN
Women	4	3	9	4
Men	6	2	6	6

Participants can then verify whether Predictive Parity is also satisfied. For women, the positive predictive value is $4/(4+3) = 0.57$, whereas for men it is $6/(6+2) = 0.75$. Consequently, Predictive Parity is violated. The contradiction emerges through exploration rather than formal proof, allowing learners to experience a simplified version of the incompatibility results discussed in the fairness literature.

Availability of Materials. To support replication, the final card templates, applicant profiles, scoring-rule sheets, facilitator instructions, and printable puzzle worksheets are released under a Creative Commons Attribution 4.0 licence at https://github.com/aro971/TheFairnessPuzzle.

4 Conclusion

Fairness has become one of the most important and widely discussed topics in contemporary Artificial Intelligence. At the same time, it remains one of the most challenging concepts to communicate effectively to non-specialist audiences. The coexistence of multiple fairness definitions, the trade-offs among competing criteria, and the presence of impossibility results make fairness difficult to reduce to simple rules or intuitive explanations.

This paper presented two unplugged activities designed to address these challenges. The first activity introduces fairness metrics through a card-based simulation of AI-supported scholarship allocation. A simple transparent scoring rule prevents the exercise from reducing AI to random choice, while the subsequent audit and token-reallocation task allows participants to explore alternative notions of fairness through direct manipulation and discussion. The second activity, the *Fairness Puzzle*, transforms fairness requirements into a collaborative constraint-satisfaction problem, enabling participants to experience the consequences of incompatible fairness criteria through guided exploration. Both activities aim to move beyond purely ethical discussions of fairness while avoiding the mathematical and technical barriers often associated with machine learning.

The activities described in this paper should be considered as a pedagogical design proposal. While the first activity builds upon a previous classroom experience that received positive feedback from students, the versions presented here have not yet been implemented or formally evaluated. Future work will therefore focus on classroom deployment, collection of qualitative and quantitative feedback, and assessment of learning outcomes. Particular attention will be devoted to investigating the effectiveness of these activities in different educational contexts, including university teaching, teacher professional development, and responsible AI training programmes for non-technical audiences.

References

1. Badaloni, S., Ferrari, C., Rodà, A.: Teaching gender knowledge and ethics in ai to stem students. In: Proceedings of the 2nd International Workshop on Education for Artificial Intelligence (edu4AI 2025), co-located with the 28th European Conference on Artificial Intelligence (ECAI 2025). CEUR Workshop Proceedings, vol. 4114. CEUR-WS.org, Bologna, Italy (2025). https://ceur-ws.org/Vol-4114/11_paper.pdf
2. Badaloni, S., Rodà, A.: Gender knowledge and artificial intelligence. In: Boella, G., D'Asaro, F.A., Dyoub, A., Primiero, G. (eds.) Proceedings of the 1st Workshop on Bias, Ethical AI, Explainability and the Role of Logic and Logic Programming (BEWARE 2022), co-located with the 21st International Conference of the Italian Association for Artificial Intelligence (AI*IA 2022). CEUR Workshop Proceedings, vol. 3319, pp. 107–112. CEUR-WS.org, Udine, Italy (2023). https://ceur-ws.org/Vol-3319/, workshop held on December 2, 2022
3. Bell, T., Rosamond, F., Casey, N.: Computer Science Unplugged and Related Projects in Math and Computer Science Popularization. In: Bodlaender, H.L., Downey, R., Fomin, F.V., Marx, D. (eds.) The Multivariate Algorithmic Revolution and Beyond. LNCS, vol. 7370, pp. 398–456. Springer, Heidelberg (2012). https://doi.org/10.1007/978-3-642-30891-8_18
4. Buolamwini, J., Gebru, T.: Gender shades: Intersectional accuracy disparities in commercial gender classification. In: Proceedings of FAT*, pp. 77–91 (2018)
5. European Parliament and Council of the European Union: Regulation (EU) 2024/1689 of the European Parliament and of the Council of 13 June 2024 laying down harmonised rules on artificial intelligence (Artificial Intelligence Act). Official Journal of the European Union, L 2024/1689 (2024)
6. Ferrara, E.: Fairness and bias in artificial intelligence: A brief survey of sources, impacts, and mitigation strategies. Science **6**(1), 3 (2024)

7. High-Level Expert Group on Artificial Intelligence: Ethics guidelines for trustworthy ai (2019). https://digital-strategy.ec.europa.eu
8. Kleinberg, J., Mullainathan, S., Raghavan, M.: Inherent trade-offs in the fair determination of risk scores. In: Proceedings of Innovations in Theoretical Computer Science (2017)
9. Mehrabi, N., Morstatter, F., Saxena, N., Lerman, K., Galstyan, A.: A survey on bias and fairness in machine learning. ACM Comput. Surv. **54**(6), 1–35 (2021)
10. Rodà, A.: The compas case: An educational journey for explaining fairness in ai-based applications. In: Calegari, R., Castillo, C., Papadopoulos, S., Søraa, R.A. (eds.) Proceedings of the 2nd Workshop on AI bias: Measurements, Mitigation, Explanation Strategies (AIMMES 2025). CEUR Workshop Proceedings, vol. 3961. CEUR-WS.org, Barcelona, Spain (2025). https://ceur-ws.org/Vol-3961/paper4.pdf

Hands-On Methods for Teaching About Generative AI

Christophe Stammet(✉)

Scienteens Lab, University of Luxembourg, Esch-sur-Alzette, Luxembourg
christophe.stammet@uni.lu

Abstract. This paper presents a set of hands-on educational activities designed to support the teaching of core concepts underlying generative artificial intelligence (AI) for both school settings and teacher training programs. While generative AI tools are increasingly present in everyday education, opportunities to develop a conceptual understanding of how these systems function remain limited. To address this gap, this paper introduces three interactive activities that aim to make central ideas of generative AI accessible without requiring any technical knowledge.

The activities focus on three key aspects: representing language as numerical values in high-dimensional spaces, understanding text generation as a probabilistic process that is prone to factual errors, and identifying how biases present in training data can influence AI-generated outputs. Each concept is translated into a structured, participatory task in which learners actively reconstruct the underlying concepts.

The paper further situates these activities within a series of workshops conducted with primary and secondary school students as well as teachers. Observations from over 50 workshops with more than 600 participants suggest that the approach is engaging, adaptable to different audiences, and presents a relevant topic to participants. These results highlight the potential of activity-based approaches to contribute to the development of AI literacy in educational contexts.

Keywords: Generative AI · Teacher Education · Hands-On Activities

1 Introduction

Generative artificial intelligence (AI) systems have rapidly become part of everyday life and have changed how the general public perceives AI tools. This shift in perception has also changed the digital environment of both students and teachers in secondary education. As a consequence, the notion of AI literacy has gained prominence as an educational goal for students and teachers alike, emphasizing that learners should not only be able to use AI tools, but also understand their underlying principles, limitations, and societal implications.

A variety of definitions for AI literacy have been proposed in publications, but research on this topic is still limited [1]. These definitions all argue for a

M. Löffler et al. (Eds.): CMSC 2026, LNCS 16951, pp. 139–149, 2027.
https://doi.org/10.1007/978-3-032-38887-2_11

combination of functional, critical, and ethical competencies that enable individuals to evaluate AI technologies, collaborate with them, and situate them in broader social contexts [2], especially since the rise of generative AI [3]. These definitions highlight that understanding core concepts of generative AI tools is as important as learning to operate them. This makes materials that enable teachers and students to understand these core concepts valuable for the growth of AI literacy in a school environment.

This paper presents such material in the form of activities from a set of workshops for primary and secondary school students, as well as non-specialist teachers that aims to address this gap by teaching important topics about generative AI through interactive, hands-on activities. These activities aim to explain three central ideas:

1. Language in modern AI systems is represented as points in a numerical embedding space, enabling models to capture similarities in meaning between words or parts of words.
2. The primary training objective of text-based generative AI is to produce text in a statistically plausible way rather than to provide factually correct information, which can lead to convincing but erroneous content.
3. Generative AI systems inherit and can amplify social biases present in their training data, resulting in stereotypical representations of people and roles in generated content.

Each idea is translated into an exercise designed for lay audiences, inviting participants to actively reconstruct the underlying principles rather than receiving them as abstract explanations.

The next three sections each introduce the theoretical foundations for one activity, followed by an overview of the activity and its required materials. The fifth section provides an overview of the workshop context in which these activities were implemented and the final section will provide a recap of the presented material.

2 How Generative AI Represents Language

This section first introduces the theoretical background behind this activity and then explains the learning objectives and procedure of the corresponding exercise.

The field of natural language processing (NLP) encompasses a variety of tasks with the goal of improving language understanding by computers, e.g. text translation or reading and writing text [4]. The first challenge lies in transforming the contents of our texts (e.g. words) into a mathematical representation that can be used algorithmically.

A technique called Word2vec [5], published in 2013, allowed for small neural networks to produce word embeddings, i.e. vectors with a few hundred dimensions, from a training on a large corpus of text. Let $d \in \mathbb{N}$ be a given dimension

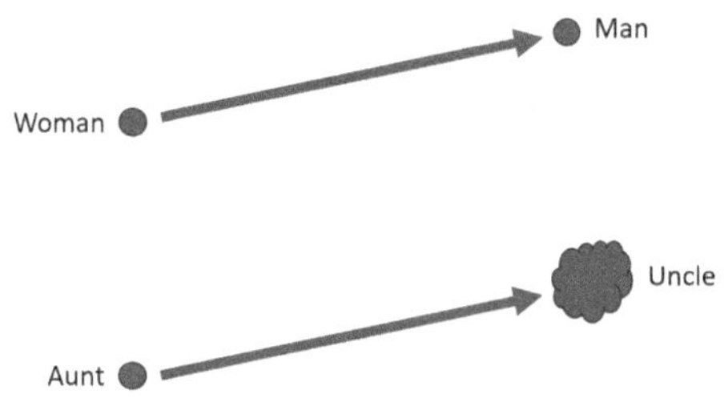

Fig. 1. Visual 2D representation of expected relations between word embeddings. Knowing the distance from "Woman" to "Man", one expects "Uncle" to be at a similar distance from the word "Aunt".

and let V be the set of all words (i.e. the vocabulary) in the training corpus C. The goal is to learn one vector $v_w \in \mathbb{R}^d$ for each word $w \in V$.

To demonstrate the coherence of the learned embedding vectors, a number of semantic tests can be performed on pairs of words. Figure 1 shows a visual 2D representation of such expected semantic and syntactic relationships between words with patterns such as "the word 'Man' is to 'Woman' what 'Uncle' should be to 'Aunt".

A similar approach to encode text as vectors is applied by large language models (LLMs), the AI models behind generative AI tools, where each dimension of the vector v_w captures patterns related to the meaning of the respective word w as learned by the model during training.

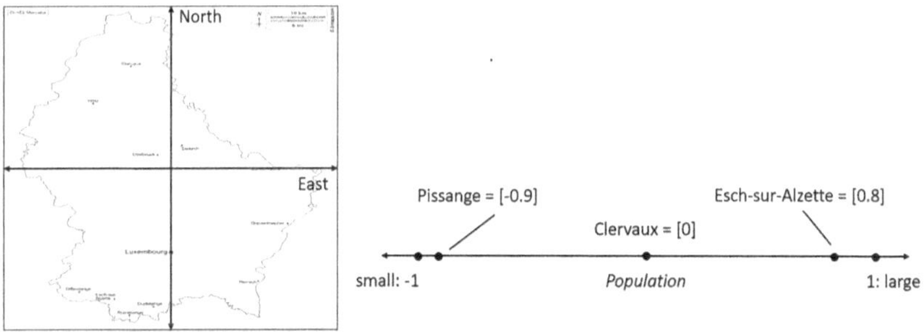

City (Token)	"East-West" Scale	"North-South" Scale	"Population" Scale
Luxembourg	0	-0.6	1
Echternach	0.6	-0.05	-0.1
Esch-sur-Alzette	-0.35	-0.9	0.8
Clervaux	-0.25	0.75	0

Fig. 2. Excerpt from slides illustrating the approach of representing token embeddings as numbers on scales of meaning.

In this exercise, embedding vectors v_w for a given word w are explained as a list of numbers used to encode the meaning of the word in human language.

Each dimension can be seen as a coordinate that describes a specific aspect of meaning. To simplify the explainability of such a vector in this exercise, it is normalized to only contain values between −1 and 1. The goal of this task is twofold:

- Understanding the concept of a list of numbers used to encode the meaning of a specific word.
- Realizing the difficulty of attributing one specific value of meaning to each coordinate.

Figure 2 illustrates the approach of explaining this idea of encoding meaning in a vector of numbers as a coordinate system, where each number places the meaning of the word on a scale that represents a certain meaning in our use of language. While individual dimensions cannot be directly interpreted in practice, this simplification helps learners grasp the idea of the embedding approach. After this introduction, the workshop participants are asked to mimic the first learning step of an LLM, where each word in a collection must be assigned a series of coordinates for a variety of meanings.

Token	"Gender" Scale	"____" Scale	"____" Scale
Child		-0,8	
Man	-1		
Gal			-0,7
Baby			
Sir			1
Bro			-1
Girl		-0,5	
Granny		1	
Woman	1		

Fig. 3. Handout to fill out by the participants. There is no unique solution and discussion among group members should be emphasized.

The handout shown in Fig. 3 is distributed to groups of two or three participants, who then work together to identify a plausible meaning for the second and third scales and to assign a numerical value between −1 and 1 to each word.

After a few minutes of work, the results from all groups are discussed together. Most groups interpret the second scale as "age", but the third scale leads to interesting discussions within the groups and a number of valid propositions, e.g. "formality", "respect", "familiarity" or "status". The values for the words tend to vary between groups for a variety of subjective reasons, which is an important part to emphasize in the discussion.

3 How Generative AI Generates Text

The original use case for the approach described in the previous section was translation of text from one human language to another. Given a source sentence in one language, the goal of the neural network architecture is to generate a target sentence in another language by encoding the source sentence into a high-dimensional vector that can be mapped by a neural network to a corresponding target vector that can be decoded to produce the translated sentence in the target language using the attention mechanism [6].

A major step towards LLMs was made by the 2017 transformer paper [7], which introduces a new architecture based on the attention mechanism called the transformer architecture. This line of work led to the first Generative Pre-Trained Transformer (GPT) model [8], designed to perform a wide range of language tasks and later adapted to more specialized use cases. After several iterations, the model GPT-3 [9] demonstrated strong language capabilities and was later optimized for conversational use through training on human dialogue. This LLM was released as the first model powering ChatGPT in November 2022.

A further development in LLM technology is Retrieval-Augmented Generation (RAG) [10], which allows a pre-trained LLM to combine its internal language capabilities with additional external sources, such as documents or datasets. A relevant application of this approach is the ability of a model to search the web and incorporate retrieved information into its answer [11].

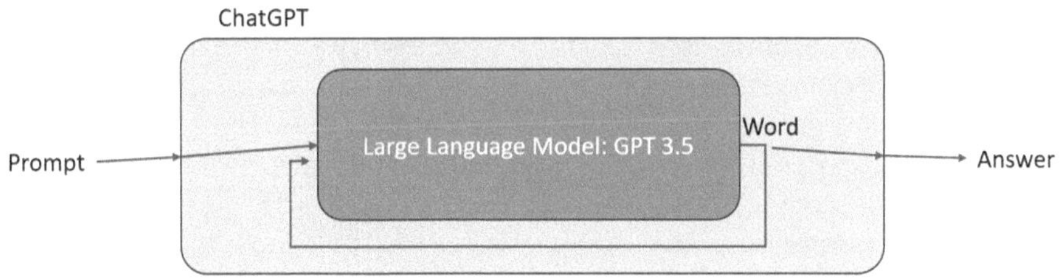

Fig. 4. Excerpt from workshop slides showing the structure of an example LLM-powered chatbot such as ChatGPT.

The key ideas are summarized in Fig. 4, which illustrates that the chatbot uses an LLM to sequentially produce words[1] to formulate an answer to the user's prompt. The focus here is on the fact that the output of the LLM is generated piece by piece, using a probabilistic choice among possible next words based on the model's training.

At the start of this workshop component, participants are invited to test two different LLMs, one with internet access and one without, on the same platform to observe differences in factual quality and writing style. We use duck.ai, as this platform does not require a log-in and promises privacy of the chats. Each participant should ask the following kinds of questions to two different LLM models in order to discuss differences and the accuracy of the replies:

- Write their own biography by providing only name and nationality.
- Translate a short piece of text from one language into another.
- Provide factual details about a topic with which the participants are familiar.

After each step, the results are discussed in class. The goal is to analyze the different answers produced by the chatbot in terms of language proficiency and factual correctness. For most prompts, participants will notice that the replies are well written but may contain factual errors. This depends on the topic of the question and on the model being used, so it is important to guide the discussion depending on the observed outputs and model differences.

4 How Generative AI Represents Bias

The previous section has shown that generative AI models produce content based on probabilistic choices, which can reduce the factual accuracy of LLM results. This is a logical consequence of such models reproducing patterns, associations, and imbalances present in their training data. The final activity shows that, as a result of this approach, model outputs can systematically favor certain demographic groups, occupations, or cultural perspectives while marginalizing others [12,13]. This makes bias an important issue when teaching about generative AI.

In LLMs, studies have found persistent stereotype bias across a range of demographic dimensions, including gender, race, nationality, religion, age, disability, and socioeconomic status [14]. These findings show that generative bias can affect both text content and the representation of people in images.

The following activity introduces the concept of bias through the following two experimental questions that the participants should investigate using generated images:

1. Is the training data biased?
2. If yes, how does this influence AI-generated output?

[1] In reality, tokens, which may be words, parts of words, symbols or similar units, are used. For simplicity, this paper treats the output of LLMs as entire words, even though the concept of tokens is explained in the full workshop.

The image generation in this exercise is done with a Stable Diffusion model [15]. To start off, the class takes a look at the dataset used for training: LAION-5B [16]. This dataset consists of over 5 billion images gathered by Common Crawl [17], a collection of images collected from openly accessible web pages.

To answer the first question, the class is separated into smaller groups of two to four participants. They first analyze the images that can be found publicly using web search for a specific gender-neutral job description[2]. The data collection uses four different search engines (Google, Bing, Ecosia, and DuckDuckGo) to ensure a larger variety in the data. Even though these images do no exactly reflect the training dataset, the patterns in public search engine results and in the contents of LAION-5B will be comparable. Each group should collect up to 50 images per search engine. The label for each picture is assigned based solely on participants' visual inference of the person on each image. An example of a partially filled-out handout can be seen in Fig. 5.

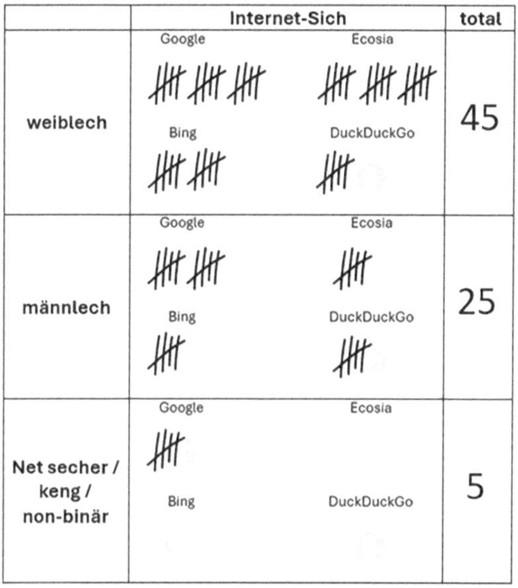

Fig. 5. An illustrative example, in Luxembourgish, of a completed handout for a group examining the genders of people in images returned for "nurse" across a variety of search engines.

After gathering the data, the next part is done with the whole group. Using a publicly available image generation platform[3], every word is given to the gen-

[2] Every group receives one specific word, for example nurse, computer scientist, or soccer player. Regardless of the language of the workshop, these words are always used in English, because they are intended to be gender-neutral.

[3] https://magicstudio.com/ai-art-generator/.

eration tool a number of times by each group[4] up to 50 generated images in total for each prompt. If applicable, the gender of the generated person is recorded in a data sheet, collected by show of hands from each group.

The results are then shown to the group and discussed in class, focusing on the fact that the distribution of the AI-generated images shows a stronger skew towards one gender than the sample from the search-engine results. An example of results from a series of workshops by a total of 73 participants can be seen in Fig. 6 for the search words "nurse" and "computer scientist", clearly showing the tendency of the AI-generated images having a stronger gender bias than the search engine results. Overall, it should be noted that this activity is mainly illustrative and aims at recognizing tendencies and not intended as a rigorous measurement of bias.

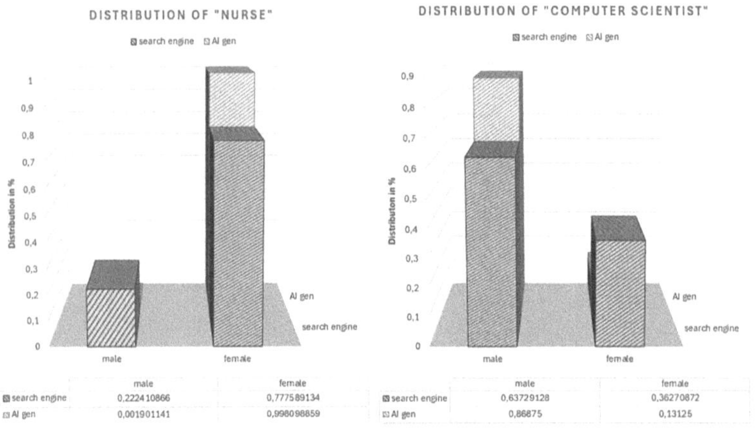

Fig. 6. Example graph of the results of the two questions showing the distributions of gender of AI-generated images and images from a search engine.

The primary takeaway to emphasize at the end of this activity is the link to AI systems and how they provide results based on their training data. An AI system produces results based on the patterns it has derived from the training data, often emphasizing the most common choice from the data. When the training data contains a bias with respect to a result, as the first experimental activity shows, the AI learns to primarily produce the more common choice, leading to the result from the second activity, where the bias from AI generated images is visibly stronger than in the images from the training data. The goal is to illustrate the presence of stereotypes in both our society and how these are amplified by AI systems and clearly state the problems of such a binary representation of stereotypes. During the discussion point of this activity, it is also important to discuss the ethical concerns of the data collection procedure in terms of privacy and consent of image data.

[4] Prompt for every job is formatted as follows: "a photo of a [job]".

5 Workshop Setting and Activities in Context

This section provides an overview of the practical contexts in which the activities presented in this paper were carried out with groups across Luxembourg. It mainly provides anecdotal evidence, as no experimental evaluation has been conducted with the material described above. In total, more than 50 workshops containing some or all of the methods presented have been organized, mainly in the following three settings:

5.1 Secondary School Classes

A total of 29 workshops were conducted with secondary school classes, for pupils aged 15 years or older, involving 499 participants in total. The methods presented in this paper were developed as part of a full three- to four-hour workshop on generative AI. The overarching goal was to provide pupils with an overview of how a ubiquitous technology such as LLM-powered chatbots functions, and what the useful applications and drawbacks of this technology are. A study has shown that 97% of students aged 17 to 30 in Luxembourg have used ChatGPT at least once [18], highlighting the importance of activities targeting AI literacy and use of LLMs.

In our general feedback form at the end of the workshop filled by all 499 participants, 91.9% of respondents indicated that they would recommend the workshop to others and 78.4% marked the relevance of the workshop's contents for their school curriculum. Some highlights from the open-ended general comment question include: *I enjoyed the workshop, it was interactive and provided insight on the subject and inspired me to look more into the topic* and *The workshop was well paced; there were no activities that I found too long or too short, and the information was presented in an interactive way.* (translated from French).

5.2 Teacher and Adult Training

A total of 14 workshops were conducted in a variety of settings, mostly for teachers in primary and secondary education, but also in several lifelong-learning settings. Each workshop lasted two to four hours and contained all the methods presented above, slightly adapted from the school-class version to better fit the target audience.

No standardized formal feedback was collected across these sessions, but teachers were positive about the workshop and the importance of the topic, with some having booked the workshop for their classes. A common point of positive feedback is that the workshop does not only present AI tools, but also explains how this technology works and presents activities that can easily be replicated in class.

5.3 Primary School Classes

Due to the age of the pupils, workshops for primary school classes, aged 11 to 13 years, focus on the bias of generative AI. The sessions for primary school are organized in collaboration with the Ministry for Gender Equality and Diversity and consist of an extended version of the method presented in Sect. 4. The workshop was offered to a total of 162 students across 11 classes in 90-minute sessions. For six of the classes, we collected feedback via a questionnaire at the end of the workshop, with strong results. On Likert-scale items (0-3 points), the workshop scored 2.46 points (*"I liked the activity"*), 2.63 (*"The concepts were clearly explained"*) and 2.5 (*"The activities were engaging and varied"*).

The answers to the open-ended questions often showed a new understanding of bias in AI and AI in general when asked about *"What new information did you learn today"*, for example *"AI can be mistaken"*, *"Not everything AI says is true"* or *"We need to ask better questions to an AI"* (translated from Luxembourgish). The teachers also showed a great appreciation of the contents of the workshop and mentioned both the importance of the topic and the difficulty for them of incorporating it into their programs without external offers such as our workshop.

6 Conclusion

This paper introduced a set of educational activities designed to teach foundational concepts of generative AI through interactive and accessible activities. By focusing on embeddings, probabilistic text generation, and bias in AI systems, the goal of the presented methods is to foster a deeper understanding of how generative AI systems work rather than simply providing tools for learners to use. For trainers or researchers interested in reproducing the activities described in this paper, the resources, materials and a worksheet for facilitators can be found on our online teaching platform: https://testit.uni.lu/resources/insidegenai/.

The activities were implemented in a variety of workshop settings, including secondary education, teacher training, and primary school contexts, demonstrating their adaptability to different age groups. Feedback collected across these settings indicates a high level of engagement and perceived relevance, highlighting the value of hands-on approaches in making complex concepts tangible and understandable.

At the same time, the results presented in this paper remain largely descriptive and based on anecdotal evidence and participant feedback. Future work could explore the integration of such activities into the formal teaching context and investigate how they support the long-term development of AI literacy.

Overall, the work underscores the importance of teaching not only how to use generative AI tools, but also how to critically understand their underlying mechanisms and limitations with practical examples that can easily be replicated in any classroom setting.

Disclosure of Interests. The author has no competing interests to declare that are relevant to the content of this article.

References

1. Laupichler, M.C., et al.: Artificial intelligence literacy in higher and adult education: A scoping literature review. Comput. Educ. Artif. Intell. **3**, 100101 (2022). https://doi.org/10.1016/j.caeai.2022.100101, ISSN 2666-920X
2. Long, D., Magerko, B.: What is AI Literacy? Competencies Des. Considerations (2020). https://doi.org/10.1145/3313831.3376727
3. Gu, X., Ericson, B.: AI Literacy in K-12 and Higher Education in the Wake of Generative AI: An Integrative Review, pp. 125–140 (2025). https://doi.org/10.1145/3702652.3744217
4. Weischedel, R., et al.: White paper on natural language processing. In Proceedings of the workshop on Speech and Natural Language, pp. 481—490. Association for Computational Linguistics, USA (1989). https://doi.org/10.3115/1075434.1075526
5. Mikolov, T., et al.: Efficient Estimation of Word Representations in Vector Space (2013). https://arxiv.org/abs/1301.3781
6. Bahdanau, D., et al.: Neural Machine Translation by Jointly Learning to Align and Translate (2016). https://arxiv.org/abs/1409.0473
7. Vaswani, A., et al.: Attention is All You Need (2017). https://arxiv.org/pdf/1706.03762.pdf
8. Radford, A., Narasimhan, K.: Improving Language Understanding by Generative Pre-Training (2018). https://api.semanticscholar.org/CorpusID:49313245
9. Brown, T., et al.: Language Models are Few-Shot Learners (2020). https://arxiv.org/abs/2005.14165
10. Lewis, P., et al.: Retrieval-Augmented Generation for Knowledge-Intensive NLP Tasks (2020). https://arxiv.org/abs/2005.11401
11. Nakano, R., et al.: WebGPT: Browser-assisted question-answering with human feedback (2021). https://arxiv.org/abs/2112.09332
12. Gallegos, I., et al.: Bias and Fairness in Large Language Models: A Survey. In: Computational Linguistics, pp. 1097–1100 (2024). https://doi.org/10.1162/coli_a_00524
13. Zhou, M., et al.: Bias in Generative AI (2024). https://arxiv.org/abs/2403.02726
14. Bai, X. et al.: Explicitly unbiased large language models still form biased associations. Proc. Natl. Acad. Sci. U.S.A. **122** (2025). https://doi.org/10.1073/pnas.2416228122
15. Rombach, R., et al.: High-Resolution Image Synthesis with Latent Diffusion Models (2022). https://arxiv.org/abs/2112.10752
16. Schuhmann, C., et al.: LAION-5B: An open large-scale dataset for training next generation image-text models (2022). https://arxiv.org/abs/2210.08402
17. Common Crawl homepage. https://commoncrawl.org/, Accessed 28 May 2026
18. BEE Secure Radar (2026). https://www.bee-secure.lu/fr/publication/bee-secure-radar/

Game-Based Learning in Mathematics and CS

GENIUS: Enhancing Mathematical Discovery through Teamwork and Play

Laura Antonelli[1](✉) and Maria Francesca Carfora[2]

[1] Istituto di Calcolo e Reti ad Alte Prestazioni, Consiglio Nazionale delle Ricerche, Naples, Italy
laura.antonelli@cnr.it

[2] Istituto per le Applicazioni del Calcolo, Consiglio Nazionale delle Ricerche, Naples, Italy
mariafrancesca.carfora@cnr.it

Abstract. GENIUS is a classroom mathematical game that combines teamwork, curiosity, and discovery. Students are divided into small groups, and each team is named after a famous mathematician: Euclid, Fibonacci, Gauss, Hypatia, Noether, and others. In this way, the game becomes not only an engaging learning experience but also an opportunity to share interesting facts about the lives and achievements of great scientists. It highlights their key contributions to mathematics while giving special attention to often overlooked figures, particularly women. At the start, each team receives a map of the human brain divided into six regions. The game is played on a floor-sized board, and teams move by rolling a die along a path of sixty squares marked with six different colours. Progression is not automatic: each team must answer a mathematical question correctly to advance. The six colours used to paint the brain regions correspond to different areas of mathematics, including logic, abstraction, numeracy, algorithms, and spatial perception, offering a varied set of challenges accessible to everyone! The first team to paint all six brain regions on its map wins, but the real goal of the game is to explore mathematical ideas in a playful and informal setting. This contribution describes the structure of the game and reflects on its effectiveness as a tool for making mathematics more engaging, collaborative, and culturally rich.

Keywords: Mathematics education · History of mathematics · Team-based learning · Women in STEM · Educational board games

1 Introduction

Well-designed mathematical tasks are invaluable educational resources and can make the difference between motivated, engaged and inspired students and those who are disengaged and unmotivated [1]. Engagement in mathematics is driven by the recognition of its creative nature and the diverse ways in which individuals can interpret and approach mathematical ideas. Rarely is there a single path to

M. Löffler et al. (Eds.): CMSC 2026, LNCS 16951, pp. 153–162, 2027.
https://doi.org/10.1007/978-3-032-38887-2_12

a solution; rather, the existence of multiple strategies and perspectives is one of the aspects that makes mathematics so compelling for learners. Although mathematical reasoning requires rigour and logical consistency, the subject comes to life when precision is combined with creativity and flexible thinking. Teachers can foster this sense of excitement through virtually any mathematical task by soliciting alternative perspectives and promoting classroom dialogue. To achieve this, it is essential to establish a classroom culture based on mutual respect, attentive listening, and the appreciation of diverse viewpoints. When students feel comfortable sharing and discussing different problem-solving strategies, classroom engagement can increase significantly, leading to richer mathematical understanding and more meaningful learning experiences [2].

From this perspective, Game-Based Learning (GBL) [3] represents a natural pedagogical choice to foster interactive engagement. While digital games dominate virtual learning environments, physical board games offer unique cognitive and social benefits by encouraging tangible, real-world interaction, collaborative decision-making, and peer-to-peer accountability. A primary benefit of GBL lies in its capacity to boost student motivation. Because games naturally revolve around challenges and rewards, they inspire students to participate actively. This hands-on involvement leads to better learning outcomes, as students retain information much more effectively when they are intensely engaged in a common task. Furthermore, GBL nurtures essential life skills, helping students practice problem-solving, teamwork, quick decision-making, and creative thinking.

Over the years, the authors' institutions, i.e. two institutes of the Italian National Research Council (CNR), IAC[1] and ICAR[2], have established extensive outreach programs aimed at promoting mathematics among young learners. These programs include educational activities and workshops for primary and secondary school students covering a broad range of mathematical topics. Furthermore, their sustained involvement in science communication events has led to the design and implementation of several informal learning experiences that engage students in mathematics outside the traditional classroom setting. Building upon this experience, this paper presents Genius, a novel physical team game designed to promote mathematics engagement in both formal classrooms and informal learning contexts. While the activity has been successfully implemented across several school workshops and public science festivals, this contribution provides the first formal account of its structural design, its integration of historical scientific narratives, and its educational potential to mitigate math anxiety. The primary objective of the proposed game is to understand that *intelligence* is a multifaceted concept encompassing a wide range of characteristics and skills. These include, among others, a strong sense of number and quantity, an accurate perception of shapes and dimensions in space, the ability to recognise patterns and formulate generalisations, and the capacity for logical and coherent reasoning. This is one of the reasons why the activity is designed to be played in teams: it allows each participant to contribute their individual

[1] Istituto per le Applicazioni del Calcolo.

[2] Istituto di Calcolo e reti ad Alte pRestazioni.

strengths while encouraging students to share and discuss their problem-solving strategies. Indeed, the ability to clearly communicate one's reasoning is a key indicator of conceptual understanding and meaningful knowledge construction.

Fig. 1. Tailored game tokens for lower-secondary school students (ages 12–15) featuring prominent mathematicians from classical antiquity to contextualise historical discoveries.

A secondary objective of the activity is to introduce students to the history of mathematics through some of its most influential scholars. The playing pieces (*tokens*) used by the teams throughout the game feature portraits of renowned mathematicians. In particular, as shown in Fig. 1, we selected mathematicians from classical antiquity for younger students (aged 12–15). For upper secondary students (aged 16–18), we chose modern and contemporary mathematicians (Fig. 2), thereby providing an opportunity to explore important milestones in the development of mathematics and scientific advances of recent centuries.

Fig. 2. Tailored game tokens for upper-secondary school students (ages 16–18) depicting modern and contemporary mathematicians, including a balanced representation of historical female scientists.

The activity was originally developed for the 2023 edition of FuturoRemoto[3] in Naples (Italy) with the topic Intelligence[4], and has been progressively refined through repeated implementations and feedback collected over the years. This iterative process has led to improvements in both the structure of the game and the educational materials supporting it. The most recent public presentation of the activity took place at the 2025 edition of Carnevale della Matematica[5] in Milan (Italy), where it was proposed as an interactive mathematics communication experience for a broad audience.

2 Game Mechanics and Design

The Genius game is played on a large circular board composed of 60 coloured squares, designed to be placed on the classroom floor and to encourage active participation and movement. The squares are distributed alternating six colours, each colour is associated with a category of questions according to the mathematical areas described in Table 1.

Table 1. Structural mapping of the Genius's mathematical domains against the OECD-PISA mathematical literacy processes and the CCSSM/NCTM standards for mathematical practice.

Color	Area of Mathematics	Main PISA Process	Related Math. Practice
●	Numeracy	Employ	Reason quantitatively
●	Iteration and generalisation	Formulate	Look for regularity in repeated reasoning
●	Logic and probability	Interpret and evaluate	Construct viable arguments
●	Measurements and statistics	Interpret and evaluate	Model with mathematics
○	Algorithms	Employ	Attend to precision
●	Spatial perception and geometry	Formulate and employ	Look for and make use of structure

The six categories of questions were designed to address different dimensions of mathematical proficiency and to engage students in a variety of logical thinking activities. Their design was inspired by the OECD Programme for

[3] FuturoRemoto is the first science festival of Europe, and is held annually in Naples (Italy), with each edition dedicated to a specific theme.

[4] The original name was in Italian, Intelligenze.

[5] Mardi Gras of Mathematics is a nationwide mathematics outreach initiative founded in 2008. Originally inspired by science blog carnivals, it combines a monthly online event with public talks, workshops, and hands-on activities.

International Student Assessment (PISA) [4], an international framework that evaluates the mathematical literacy of 15-year-old students, focusing on their ability to formulate, employ, and interpret mathematics in real-world contexts. In addition, the categories reflect the mathematical practices described in the NCTM[6] standards and in the Common Core framework [5,6] (see Table 1). The game's challenges are derived from reputable national and international mathematics competitions, such as the Mathematical Olympiads and the Kangaroo Math Competition, as well as high-quality online resource repositories including NRICH[7]. These curated tasks were systematically categorised and adapted to align with the target players' age groups. The pedagogical adaptation focused on restructuring the original problems to make them accessible to students of average mathematical ability, ensuring they could be solved successfully within the strict time constraints required to maintain a fast-paced and engaging gameplay experience. Participants are organised into small teams, each identified by the name of a famous mathematician (see Figs. 1 and 2). According to the number of players, it is possible to organise a number of teams from two to six. Every team receives a playing token and a score sheet depicting a human brain (see Fig. 3). Progress in the game is represented by gradually colouring different brain areas whenever the team provides a correct answer, thus offering a visual representation of knowledge acquisition.

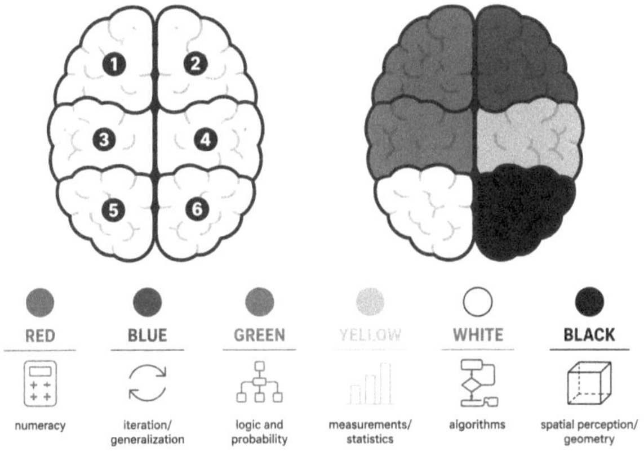

Fig. 3. The score sheet representing a human brain map, where teams track their conceptual progression by colouring regions associated with the six mathematical categories described in Table 1.

At each turn, a team rolls a die and moves its token accordingly around the board. The colour of the square determines the category of the question

[6] Principles and Standards for School Mathematics of the National Council of Teachers of Mathematics (NCTM).

[7] https://nrich.maths.org/.

to be answered, and an example of a colour-based question for each category is shown in Fig. 4. Team members are encouraged to discuss possible solutions collaboratively; however, to promote the active involvement of all participants, a different student is required to present the team's final answer at each turn.

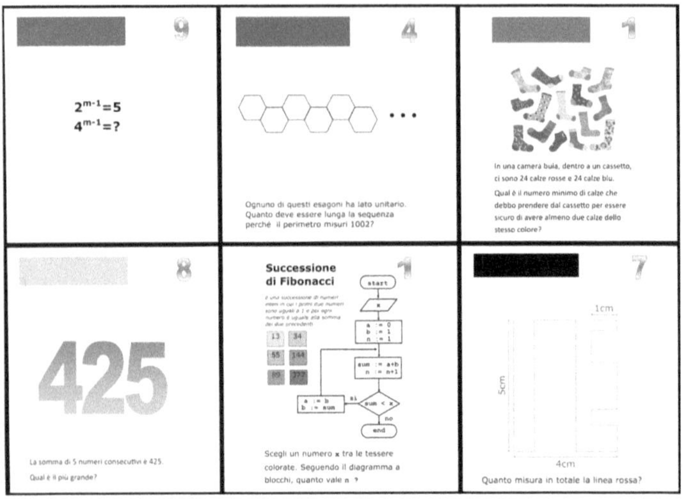

Fig. 4. Representative samples of question cards for each mathematical category (see Table 1), adapted from national and international competitions to fit fast-paced gameplay constraints.

Advancing in the game through the different question categories fosters several Mathematical Practices, including making sense of problems and persevering in solving them, reasoning abstractly and quantitatively, constructing viable arguments, critiquing the reasoning of others, and recognising patterns and structures. The collaborative nature of GENIUS further supports mathematical communication and collective knowledge building, as students are required to discuss possible strategies and justify their answers before presenting them to the class.

The combination of physical movement, collaborative problem solving, and turn-taking mechanisms is intended to foster engagement, mathematical communication, and equitable participation within the group. Furthermore, the visual progression system provides immediate feedback and reinforces students' sense of achievement throughout the activity. This mix of activities makes GENIUS a game that promotes a richer conception of mathematics learning, in which procedural fluency is integrated with conceptual understanding and mathematical thinking. Throughout the gameplay, specific board spaces are dedicated to the historical figure depicted on the player's token. Landing on these character-specific squares triggers a dedicated storytelling and quiz phase, designed to deepen engagement and historical contextualization. Players are challenged by accessible questions about the mathematician's life, ranging from their chronology and notable personal anecdotes to the groundbreaking achievements that defined their career. Successfully answering these questions serves as a core progression mechanic, allowing players to advance further along the board. This integration of narrative and trivia not only incentivises active learning but also transforms historical context into a rewarding gameplay element.

3 Field Evaluation and Case Studies

The GENIUS activity has been implemented in both formal and informal educational settings, involving students of different age groups as well as people of the general public. In all cases, participants worked in small teams of three to five members, encouraging discussion, collaboration, and the sharing of problem-solving strategies.

3.1 Public Engagement and Informal Learning Contexts

As detailed in Sect. 2, the tailored sets of question cards and age-based character tokens effectively triggered peer discussions. Typical sessions lasted between 40 and 60 min. Following a brief rules introduction, teams competed to solve mathematical riddles scattered across the floor-sized board. Facilitators actively encouraged participants to explain and justify their reasoning, thereby promoting mathematical communication and peer-to-peer learning. Informal observations revealed high levels of engagement, particularly when teams compared alternative solution strategies. This initial pilot proved critical for iterative refinement, allowing the authors to recalibrate challenges that were identified as either too simplistic or overly complex. The activity reached its final, consolidated design for the 2025 edition of the CARNEVALE DELLA MATEMATICA DAL VIVO. Here, groups of high school students participated in structured 50-minute sessions followed by a reflective post-game debriefing. Notably, the most enthusiastic and positive feedback came from students attending humanities and artistic high schools. This outcome strongly confirms the validity of the tool, demonstrating its capacity to mitigate math anxiety and foster a genuinely positive attitude toward mathematics even among student populations traditionally less aligned with STEM disciplines (Fig. 5).

3.2 Formal Classroom Implementation

From 2024 to 2026, the activity was also proposed to approximately 20 middle and early high school classes during regular school hours, integrated within dedicated STEM competence enhancement programs. Initial student hesitation was rapidly overcome, largely due to the physical rearrangement of the classroom space required to construct the game board and the autonomy granted to students to form their own teams. Notably, selecting a team token triggered

Fig. 5. Interactive gameplay sessions involving floor-sized board dynamics during the public outreach pilot phase at the FUTUROREMOTO 2023 science festival.

an unplanned preliminary debate regarding the historical mathematicians represented. This choice fostered a meaningful deep dive into lesser-known female figures in mathematics, addressing gender representation in science from the very outset of the activity. The teachers' involvement provided critical insights into the orchestration of game-based learning. In several instances, educators immediately embraced an active role as facilitators, supporting struggling groups and encouraging inclusive participation. Conversely, other teachers experienced greater friction in shifting from traditional instruction to an active-learning stance, highlighting varying levels of comfort in monitoring peer-group processes and dynamically mediating student interactions during gameplay. A post-game debriefing session with students served to highlight both the strengths and constraints of the activity. First, students' feedback regarding the perceived difficulty of the trivia and challenges offered invaluable data for iterative game design, shedding light on how to reformulate specific question cards. Second, students actively reflected on their internal team dynamics, noting how peer-to-peer discussion clarified their mathematical reasoning and enhanced their comprehension of complex problems. Ultimately, across all participating classes, students universally reported that the activity heightened their curiosity and fostered a significantly more positive attitude toward problem-solving and mathematics as a discipline.

4 Discussion and Future Work

The preliminary implementations of GENIUS offer encouraging indications regarding its potential as a tool for promoting active engagement in mathematics education. Although a systematic quantitative evaluation has not yet been conducted, informal observations and feedback collected from both students and educators demonstrate high levels of enthusiasm and sustained motivation throughout the activity.

A prominent insight emerging from the fieldwork concerns the impact of collaborative dynamics. The game structure prompts students to debate potential solutions, compare cognitive strategies, and negotiate shared answers. This collaborative dimension facilitates peer-to-peer learning and lowers the barrier to entry for students who traditionally remain disengaged during conventional mathematics lessons. Furthermore, the mechanic requiring a different team member to deliver the final answer each turn ensures a more equitable distribution of accountability and active participation. Crucially, the framing of competition within the gameplay. Rather than inducing performance anxiety, the competitive component remains playful and low-stakes, as progression relies on a balanced mix of mathematical competence and stochastic elements, such as dice rolls. This equilibrium shifts the focus from performance to exploration, creating a safe learning environment where mistakes are reframed as constructive iterations rather than failures. Consequently, students appeared intrinsically motivated to tackle complex questions and frequently initiated spontaneous discussions about alternative solution strategies even after a turn had concluded. Feedback gathered during these pilot phases has also been instrumental for material optimisation. Students' responses and teachers' insights provided empirical data to refine the calibration and classification of the question cards based on their perceived difficulty and pedagogical effectiveness. This iterative revision process has significantly enhanced the balance across the different mathematical domains and challenge tiers represented on the board. At present, the evidence supporting the efficacy of GENIUS is primarily qualitative and perceptual. Future work will therefore focus on implementing a systematic evaluation framework. We aim to collect structured data through validated questionnaires and assessment rubrics administered to both students and teachers post-activity. Additionally, video recordings of gameplay sessions will be integrated to conduct qualitative discourse analysis, yielding deeper insights into the dialogic relationships and collaborative reasoning that emerge during group problem-solving. Subsequent studies will also investigate the alignment between game engagement and the development of specific mathematical competencies, particularly those operationalised by the OECD-PISA framework and the Mathematical Practices highlighted in contemporary educational research. Such analyses will establish a robust empirical foundation to assess the long-term educational impact of GENIUS and to guide future iterations of its core game design. Future work will also include the online publication of the game rules, board templates, question cards, and supporting materials, allowing secondary schools to freely adopt and adapt the activity in their own educational contexts. The availability of the materials as open educational resources may facilitate their reuse, adaptation, and further development by teachers and researchers interested in innovative approaches to mathematics education.

5 Conclusion

GENIUS reimagines the mathematics classroom not as a place for passive instruction, but as an open ecosystem of exploration, low-stakes competition, and horizontal dialogue. By merging rigorous curriculum-aligned challenges with physical movement and historical narrative, the game successfully detaches mathematics from the stigma of anxiety and absolute abstraction. For the mathematics communication community, this project demonstrates that shifting from *performing math* to *playing math* bridges the gap between STEM disciplines and students traditionally disengaged from scientific fields, such as those in the humanities. Ultimately, GENIUS offers a scalable open-access framework that empowers educators to transform procedural fluency into shared, culturally rich conceptual understanding, proving that mathematical literacy is inherently a collective adventure.

Acknowledgments. The authors would like to express their gratitude towards CITTÁ DELLA SCIENZA in Naples (Italy), where the board game was first presented during the edition of FUTUROREMOTO 2023 – INTELLIGENZE. ChatGPT (OpenAI) was used for language editing and stylistic refinement of the manuscript, as well as for generating Fig. 3. The authors take full responsibility for the content of the paper.

Disclosure of Interests. The authors have no competing interests to declare that are relevant to the content of this article.

References

1. Boaler, J.: Mathematical Mindsets: Unleashing Students' Potential Through Creative Math, Inspiring Messages and Innovative Teaching. Jossey-Bass/Wiley (2016)
2. Klang, N., Karlsson, N., Kilborn, W., Eriksson, P., Karlberg, M.: Mathematical problem-solving through cooperative learning - the importance of peer acceptance and friendships. Front. Educ. **6**, 710296 (2021)
3. Plass, J.L., Homer, B.D., Kinzer, C.K.: Foundations of game-based learning. Educ. Psychol. **50**(4), 258–283 (2015)
4. OECD. PISA 2022 Assessment and Analytical Framework. OECD Publishing, Paris (2023)
5. National Council of Teachers of Mathematics: Principles and Standards for School Mathematics. NCTM, Reston, VA (2000)
6. National Governors Association Center for Best Practices and Council of Chief State School Officers. Common Core State Standards for Mathematics. NGA Center and CCSSO, Washington, DC (2010)

Playing the Odds: Bids, Beliefs and Probabilistic Reasoning in Indian Classrooms

Neha Garg and Jyothi Krishnan(✉)

Centre for Creative Learning, IIT Gandhinagar, Gandhinagar, India
{neha.g,jyothi.k}@iitgn.ac.in

Abstract. We describe a four-session, game-based probability module, facilitated online with middle school girls at government residential institutions serving disadvantaged communities across India. Through two dice games, students considered short-run variation, unequal likelihoods, and the different numbers of ordered outcomes underlying the sums of two dice. We describe how facilitated gameplay made students' probabilistic intuitions visible and how some groups worked to connect ordered outcomes, distributions, symmetry, and payouts. Two groups subsequently developed three-dice versions of a betting game by counting outcomes, using symmetry, and assigning larger payouts to less likely sums. Brief observations from sessions with senior secondary students and teachers suggest that similar intuitions may persist after formal instruction. The account describes how games, representations, and facilitation were combined to create opportunities for students to articulate and examine probabilistic reasoning.

Keywords: probability education · game-based learning · combinatorial reasoning · ordered outcomes · under-resourced settings

1 Introduction

Ideas about probability arise before students encounter the subject formally. Students notice variation and make judgements about likelihood through everyday experiences: how long they wait to roll a six in *Ludo*[1], how often a bus is late, or how often a coin toss determines which cricket team bats first. These intuitions, however, do not always align with formal probability. One common difficulty is the *equiprobability bias* [1]: treating distinct outcomes as equally likely even when they correspond to different numbers of elementary outcomes.

In addition, learners frequently judge the sums 2–12 on two dice to be equally probable, overlooking that 7 arises from six ordered pairs while 2 and 12 arise from only one each [2]. Reasoning about such compound events requires learners [3,4] to determine when order matters and enumerate the possible outcomes

[1] A board game, popular in India, in which a player needs to roll a six to start.

M. Löffler et al. (Eds.): CMSC 2026, LNCS 16951, pp. 163–177, 2027.
https://doi.org/10.1007/978-3-032-38887-2_13

without omission or repetition. Research suggests [5] that understanding two-dice sample spaces develops through intermediate forms: students may recognise that a sum can occur in several ways before distinguishing ordered outcomes or coordinating them into a complete distribution. Such intuitions may persist alongside formal knowledge. Knowing a probability rule does not ensure that it will be applied when interpreting a short sequence, judging a random device, or making a game decision. Fischbein [6] argued that beliefs about chance rooted in experience are not necessarily displaced by instruction, while Watson and Moritz [7] found that students' stated beliefs about dice fairness did not always correspond to the strategies they used to investigate it. A related difficulty concerns short-run variation. Learners may focus on what should happen on the next trial rather than on the distribution expected over many trials, an orientation Konold [8] terms the *outcome approach.*

This paper presents a practitioner account of a four-session, game-based probability module with middle-school girls attending Kasturba Gandhi Balika Vidyalayas (KGBVs) in India. Before formal explanations were introduced, students decided whether to roll again, where to bid, and how to interpret payouts. Pictographs, simulations, distinguishable dice, and outcome grids were then used to examine the reasoning that emerged. We focus on students' judgements about chance, the movement from labelled sums to ordered outcomes, the relationship between theoretical probability and short-run variation, and two groups' attempts to extend a two-dice game to three dice. The aim is not to measure learning gains, but to describe how students reasoned during these activities.

2 Context, Module Design, and Sources of Evidence

2.1 Programme Context

The module formed part of a large-scale online STEM programme involving KGBVs, government-run residential institutions established for girls from disadvantaged and historically marginalised communities across India. Each KGBV serves about sixty middle-school girls, although enrolment and institutional arrangements vary across states. Students may attend nearby government schools or receive instruction within the residential institution.

The biweekly math and science programme enrolled 1346 KGBVs across six states and one union territory, although participation varied across sessions. An average of 515 KGBVs attended this module live (see Table 2 in Appendix A for details), with recordings made available for later viewing. Students joined the sessions after school, gathering in a common room and working with inexpensive materials that could be improvised locally (Fig. 1). The sessions were conducted predominantly in Hindi.

Ethical approval for the programme was obtained from the IIT Gandhinagar Institute Ethics Committee. Assent and consent were obtained as required. Student names used in the paper are pseudonyms, and photographs and artefacts are reproduced with permission.

Fig. 1. Views of the Sessions

2.2 A Four-Session Game Module

The module used two dice-based games, *Watch Your Step* and *Dicey Dice.* We use games since they make likelihood consequential: a bid or a decision to roll again forces a probabilistic judgement that a definition-first lesson can leave implicit. The four-session sequence was designed so that these judgements surfaced in play before representations – pictographs, simulations, distinguishable dice, outcome grids – were introduced to examine them. The facilitator was also an active participant in the episodes: modelling reasoning aloud, selecting representations, deferring some answers, and repeatedly asking students to explain or extend their ideas. Table 1 summarises the instructional sequence.

Table 1. Overview of the Four-Session Probability Module

Game and sessions	Focus
Sessions 1–2 (*WYS*)	Single-die likelihood, short-run variation, fairness
Sessions 3–4 (*DD*)	Two-dice sums, ordered outcomes, payouts

WYS: *Watch Your Step*; *DD*: *Dicey Dice.*

2.3 Sources of Evidence

This experience report is grounded in records generated through the programme: poll responses, chat logs, homework submissions, photographs/videos of student work shared by schools and recordings of the live sessions. Homework consisted of worksheets (see Table 3 in Appendix A). Responses were collected via Google Forms. Polls and worksheets were submitted collectively by a participating school rather than by individual students. In addition to responding through polls, students held up fingers to indicate numerical answers. Detailed dialogue came from the much smaller subset of schools that volunteered to be spotlighted and interacted live or submitted explanatory artefacts (Appendix B). The facilitator invited different schools to participate publicly, with 18 distinct KGBVs spotlighted across Sessions 3 and 4.

We reviewed these records for episodes in which students expressed beliefs about likelihood, revised a strategy, connected labelled outcomes to ordered outcomes, connected probability to payouts, interpreted a distribution, or extended the game structure. The episodes below were selected because they provide particularly clear instances of these forms of reasoning. They should not be read as representative of all participating schools or as evidence of measured change across the cohort.

3 Rolling One Die: *Watch Your Step*

Watch Your Step, designed by Mike Petty, is a two- to four-player game played with a single die on a board of nine "rocks" (Fig. 2a). On each turn, a player rolls the die and moves that many steps forward. If they remain on the rocks, they may stop or roll again; the turn continues until they stop or fall into the water. Stopping on rock 6, 7, or 8 scores one point, while stopping on rock 9 scores three. Overshooting rock 9 sends the player into the water and costs one point. The player furthest along the path at the end of the round receives two additional points. Once every player has taken a turn, all tokens are returned to the starting position for a new round. The game ends at the end of the round in which a player reaches 12 points.

3.1 'Lucky' Numbers and Small Samples

The module opened with a poll, "Which number is most likely on a throw of a die?" Of the 176 KGBVs that responded, only 86 selected the option that all six outcomes of a fair die were equally likely. The remainder favoured particular numbers, with 57 selecting six as the most likely outcome. We deferred discussion of the poll, taught students the game, and asked them to play ten rounds of *Watch Your Step* as homework, record their rolls, and send photographs of their results (Fig. 2b). When the pictographs were compared in Session 2, the schools' small samples differed: one showed 1 most often, another favoured 4 and 5, and a third appeared to support the initial preference for 6.

This disagreement became the starting point for discussion: why had schools rolling apparently similar dice obtained different results? A repeated-roll simulation displayed substantial variation in its early stages but became increasingly

(a) Game Board (b) A Student Tracks Rolls

Fig. 2. From *Watch Your Step* Sessions

uniform as it was extended to 1000 rolls, illustrating the contrast between short-run variation and long-run regularity.

We also examined how the physical structure and labelling of a die affect fairness. Drawing on an earlier discussion of Platonic solids, we discussed congruent faces and geometric symmetry as features of a plausible fair randomiser. They reasoned that an irregular solid with unequal faces would not behave fairly and immediately identified 4 as the most likely outcome on a die labelled 1, 2, 3, 4, 4, 4. These exchanges connected empirical results, physical symmetry, and label frequency as different forms of evidence for judging dice fairness [7].

3.2 To Stop or to Roll Again?

The recurring decision in *Watch Your Step* was whether to roll again: some groups stopped as soon as they reached a scoring position, while others continued. As a spotlighted group chose to roll again the facilitator reminded them that they might fall. One student replied, *"Kya pata jeet jaayein?"* ("Who knows, we might win?"). Another asked, *"Ma'am, kya pata pehla paasa girane wala aa jaaye?"* ("What if the first roll makes us fall?"). While the early focus was on the favourability of the next roll, by the end of Session 2 a student had formulated a more general strategy: roll again whenever the number of scoring outcomes was at least as large as the number of falling outcomes–replacing attention to an isolated possibility with a comparison across the available outcomes.

3.3 An Unexpected Hurdle

Not every difficulty was probabilistic. Asked to total each round's score, many students added the positive points and did nothing with the negative ones; they seemed unable to recognise '−1' as a number. The same trouble appeared in the homework. This arithmetic obstacle is worth recording, because especially in an under-resourced setting, the prerequisite skill can gate a lesson as surely as the target concept does. Thus in this part of the first session, difficulty with signed integers became a more immediate obstacle than probabilistic reasoning itself. To accommodate this, we moved tokens forward and backward to keep score and later added a tracker to the board.

4 *Dicey Dice*: Betting on Two-Dice Sums

Dicey Dice is adapted from a two-dice casino-style betting game. The board contains fifteen spaces: the sums 2–12 and four category spaces – Odd, Even, Under 7, and Over 7. Each player begins with ₹10,000 and places one to three bids on any of the spaces, in multiples of ₹1,000. The dice are rolled and payouts are made based on the sum of the two dice. Winning bids are paid at the rate printed on the corresponding space, and losing bids are forfeited. After ten rounds, the player with the most money wins. The board we used is in Fig. 3a.

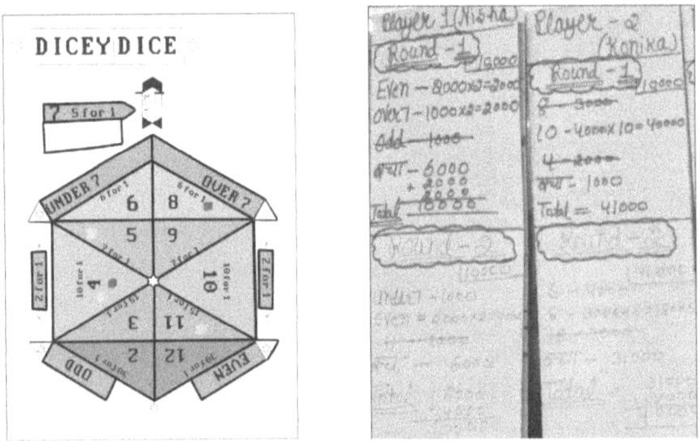

(a) Game Board (b) Students Keep Score

Fig. 3. From *Dicey Dice* Sessions

4.1 From Bids to Ordered Outcomes

Session 3 began with an introduction to *Dicey Dice* and keeping score (Fig. 3b). Schools were spotlighted in turn to place their bids before the facilitator rolled the dice. Early bids were often guided by preference, payout size, or hope rather than by probabilistic reasoning. A group from a KGBV in Himachal Pradesh had been winning repeatedly. Asked to explain, the group described its strategy: *"Ma'am, Bhagwan ka naam leke!"* ("Ma'am, by invoking God's name!"). Noticing that one of the two was assured, several groups bid simultaneously on Odd and Even. The facilitator modelled the financial consequence: under the game's payout structure, equal bids on both spaces guarantee one winning category but leave the balance unchanged. One group later produced an analogous Under 7 and Over 7 strategy, not yet recognising the contribution of 7.

An online dice display (seen in Fig. 4a) was used to present a predetermined sequence of outcomes. This ensured that students encountered repeated sums arising from different ordered pairs, particularly several appearances of the sum 6. Session 4 opened with a poll and, not unexpectedly, the majority of responding schools selected the response that all sums from 2 to 12 were equally likely. Rather than directly tackle the poll, the facilitator asked what they noticed about the *ways* in which 6 had arisen. Sohni and Heer, from KGBV Loni, pointed out that the same sum had appeared as both $(4, 2)$ and $(1, 5)$. The discussion then moved from whether a sum appeared to how many ordered outcomes could produce it. Everyone agreed that 2 could arise only as $(1, 1)$. The sum 3 divided the responses: some groups counted "2 and 1" as a single option, while others distinguished $(1, 2)$ from $(2, 1)$. In the following exchange with Mira from KGBV Gosaiganj, we used differently coloured dice to make the order visible:

Neha: In how many ways can you get 3?
Mira: One way, ma'am; 2 and 1.
Neha: You said 2 on one die and 1 on the other. On which die is the 2, and on which is the 1? *(shows 2 on the red die and 1 on the blue die)*
Mira: On either of them.
Neha: What if the red die shows 1 and the blue die shows 2? Is that valid?
Mira: Yes, ma'am.
Neha: So how many ways?
Mira: Two ways.

As a student enumerated pairs summing to 8, the facilitator suggested $(1, 7)$, prompting her to object: *'Ma'am, 1 aur 7 kaise hoga? Usmein toh 6 tak hi hain"* ('Ma'am, how can there be a 1 and a 7? The die only goes up to 6."). We used this opening to move toward a systematic enumeration.

4.2 From Outcome Counts to a Distribution

Students listed the ordered pairs producing each sum on a two-dice outcome grid (Fig. 4b). As the grid filled, the number of ordered pairs increased from one to six for sums from 2 to 7, then decreased symmetrically. One group described the resulting shape as a "triangle" or a "Christmas tree," and eighteen schools noted in the chat that 7 had the greatest number of ordered pairs. The representation shifted attention from the eleven possible sums to the thirty-six equally likely ordered outcomes that produced them.

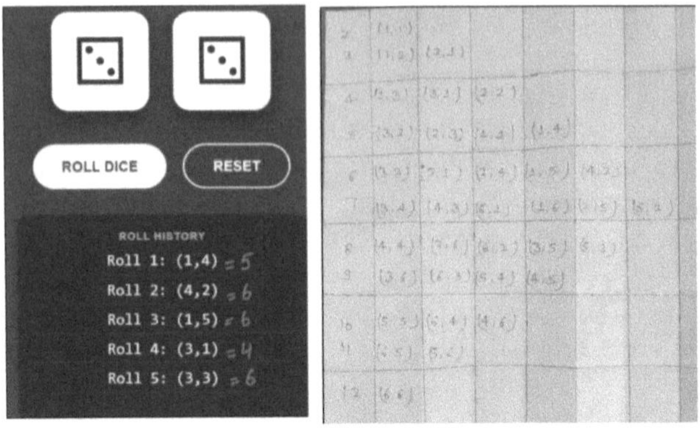

(a) Rigged Outcomes (b) Seeing the Distribution

Fig. 4. Rolling Two Dice

4.3 Relating Probabilities to Payouts

The payout structure of *Dicey Dice* encoded the unequal distribution of sums. Asked why 2 and 12 carried the largest payouts, a student from KGBV Kaudihar, Uttar Pradesh, replied, *"Paasey ki ek hi baar chance hai aane ki"* ("The dice have only one chance of coming that way"). Although informal, the response connected the single ordered outcome producing each extreme sum with its larger payout. Worksheet responses were broadly consistent with these exchanges. By the end of Session 4, 184 of 317 responding schools recognised that equal bids on Odd and Even yielded no net gain. 208 of 317 rejected the claim that even sums were more likely simply because there are six even totals and five odd totals. 167 of 316 also correctly compared the probability of a new Prime category with that of 7. These responses do not establish learning gains, but they are consistent with the reasoning observed in the live exchanges: counting the ordered outcomes within each category rather than merely counting category labels.

5 The Three-Dice Challenge

Near the end of Session 4, after connecting the two-dice distribution to the board's payouts, students from KGBV Loni spontaneously asked *"Ma'am ji, agar hum yeh teen dice se khelein?"* ("Ma'am, what if we play this with three dice?"). Rather than answer the question, we returned it to the students as a design challenge: create a version of *Dicey Dice* for three dice—an approach in which learners invent or modify games to deepen probabilistic reasoning [9]. We asked them to decide which sums the board should contain, how the payouts should be assigned, and how the board might be organised. This extension substantially increased the mathematical demands of the task. With three dice, the sample space expands to 216 ordered outcomes, distributed across the sixteen sums from 3 to 18, with a joint maximum at 10 and 11.

5.1 A Productive, Though Incomplete Use of Symmetry

A day later, the KGBV in Bhatiya, Gujarat, submitted a game board (in Fig. 5a) through the programme's WhatsApp group. In a follow-up Zoom call, the students explained that they had begun enumerating outcomes from a sum of 3 but stopped at 9 because the counting had become unwieldy. Their count for 9 was incomplete—approximately fifteen or sixteen outcomes rather than the correct twenty-five. They nevertheless attempted to complete the remaining side of the board by using the symmetry they had observed in the two-dice distribution.

Because their enumeration stopped before the centre of the three-dice distribution, the resulting board placed the apparent peak at 9 rather than at 10 and 11. The board was therefore only a partial reconstruction. The students recognised that the outcome counts for higher and lower sums should be symmetric, but applied this pattern before locating the centre of symmetry. This difficulty is consistent with Watson and Kelly's [5] finding that understanding demonstrated with two dice may not transfer intact to a more complex three-dice task.

5.2 Coordinating Outcome Counts, Symmetry, and Payouts

When we contacted the school that had originally posed the question, students from Loni, Uttar Pradesh, submitted a completed board (see Fig. 5b) the following day. We arranged a call to understand the group's reasoning. A student stated the sample-space structure directly: one die has 6^1 outcomes, two dice have $6^2 = 36$, and three dice have $6^3 = 216$. The group had worked out the number of ordered outcomes for successive sums: one way to make 3, three ways to make 4, six ways to make 5, and so on. They then assigned larger payouts to sums with fewer outcomes and smaller payouts to sums with more outcomes. The board was organised symmetrically, with the lower sums on one side and the corresponding higher sums on the other. The students used reflection to complete the far side after counting the near side rather than enumerating every sum independently.

We later asked the Loni group to send a video explaining their board design. The students were leaving for vacation and said that they would send one after returning. When we expressed concern that they might forget, they replied, *"Ma'am ji, yeh toh humne khud banaya hai, ise kaise bhul sakte hain?"* ("Ma'am, this is something we made ourselves—how could we possibly forget it?"). The explanatory video arrived after the winter vacation. Their willingness to undertake the lengthy enumeration required by the redesign, and to return later to explain it, suggests sustained interest in the mathematical work as well as a strong sense of ownership of the resulting artefact. The mathematical significance of the extension lies in the direction of reasoning. During the original game, students could move from the board's payouts toward an inference about which sums were likely. In the redesign, the direction was reversed: the groups began with ordered outcomes, constructed or approximated a distribution, and used that distribution to assign payouts.

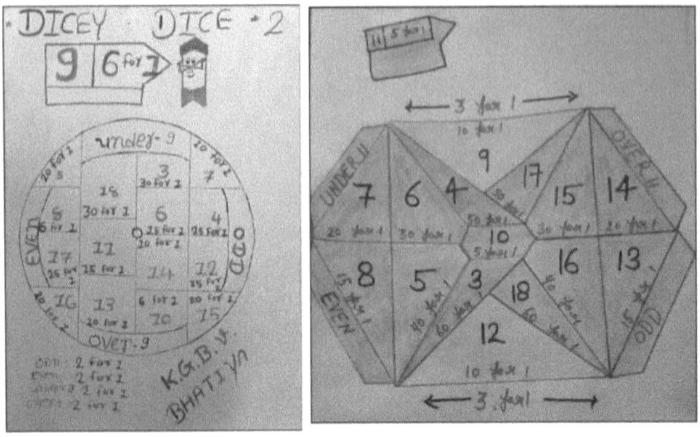

(a) KGBV Bhatiya (b) KGBV Loni

Fig. 5. Student-Designed Game Boards

6 *Dicey Dice* in Other Contexts

6.1 After Curricular Exposure

A group of six Grade 11 girls from a well-resourced private school played *Dicey Dice*. While probability formed part of their curriculum, initial gameplay suggested this knowledge was not spontaneously applied. One student remarked that "the probability of 12 coming is small," yet the group asserted that all sums from 2 to 12 were equally likely, estimating $P(7)$ as $1/11$ and judging even sums more likely than odd because the range held six even values and five odd. The facilitator asked them to enumerate ways of forming particular sums. The students recognised that 7 could arise as $(1, 6)$, $(2, 5)$, $(3, 4)$ and their reverses, while 2 and 12 each had only one ordered outcome. After some difficulty identifying the denominator, one student reasoned that two dice produce 36 ordered outcomes, and the group obtained $P(2) = P(12) = 1/36$ and $P(7) = 6/36$. Extending this to the category spaces, they initially assigned $14/36$ to both Under 7 and Over 7, then revised to $15/36$ when prompted to check that the two categories and 7 covered the space. Using probabilities and payouts, they concluded that bids on Odd, Even, and 7 offered a larger possible gain than Under 7, Over 7, and 7.

6.2 With Practising Teachers

Dicey Dice was used in several professional development workshops with practising teachers, where similar judgements surfaced in discussion. When asked which sum from 2 to 12 was most likely, participants sometimes began by treating the eleven labelled sums as equally likely and, when prompted to reconsider, identified 12 as more likely than the other sums.

7 Discussion

Game Decisions Made Probability Consequential. The games required students to act on judgements of likelihood. In *Watch Your Step*, they had to decide whether another roll was worth the risk of falling. In *Dicey Dice*, they had to judge whether a large payout compensated for a rare outcome and whether combinations of bids could produce a gain. These choices elicited reasoning that a definition-first lesson might not: preferences for particular outcomes, attention to isolated possibilities, attraction to large payouts, and the assumption that labelled sums are equally likely.

Representations Changed What Could be Seen. Pictographs and repeated-roll simulations made short-run variation visible. Coloured dice distinguished $(1, 2)$ from $(2, 1)$, while the ordered-outcome grid revealed the unequal distribution of sums and its symmetry. These representations did more than illustrate conclusions: they changed the objects available for comparison—faces rather than labels, ordered outcomes rather than sums, and distributions rather than isolated results.

Facilitation Kept the Reasoning with the Students. The key moments depended on facilitator moves such as deferring discussion of the opening poll, comparing pictographs from different schools, modelling strategic reasoning, presenting a predetermined sequence of sums, distinguishing the dice by colour, and returning the three-dice question as a design challenge. The games created a need for probabilistic reasoning; facilitation and representation made that reasoning visible and open to examination.

Formal Instruction Did Not Guarantee Spontaneous Application. The Grade 11 episode suggests that, despite prior exposure to probability, students did not spontaneously construct the thirty-six ordered outcomes while playing and often treated the eleven labelled sums as equally likely. Informal observations from workshops with practising teachers also showed similar judgements. Although these observations do not establish how widespread either judgement was, they suggest that participants continued to reason from the labelled sums rather than from the ordered outcomes that produced them. This is consistent with Fischbein's argument that intuitive beliefs about chance may coexist with formal knowledge [6]. The difficulty was therefore not simply the absence of a formula; whether an underlying combinatorial structure is recognised and applied can depend strongly on how the task is represented [10].

Acknowledgments. This work was carried out as part of the Curiosity Programme at the Centre for Creative Learning, IIT Gandhinagar. The authors thank the Centre for its institutional support, and Samagra Shiksha, the participating KGBVs, and the teachers and students from Uttar Pradesh, Gujarat, Himachal Pradesh, Punjab, Tripura, Rajasthan, and Daman and Diu whose engagement made this work possible.

Disclosure of Interests. The authors have no competing interests to declare that are relevant to the content of this article.

A Data Across the Four Sessions

Table 2. Participation Averaged Across the Four Probability Sessions

Measure	Average
Schools attending Zoom live	243
Schools attending YT live	272
Schools watching recorded sessions	34
Students attending	46,715
Worksheet submissions (schools)	331
Schools submitting photographs and videos	121
Schools interacting live	15
Zoom Poll responses	140

A.1 In-Session Polls

Watch Your Step – Session 1

1. Which number is most likely on a die?
2. On your turn you roll 2, 4 and 1. Which rock are you on and what is your score?
3. Who will get a bonus in this round? (picture given)
4. Which number on the die scores on the first turn?

Watch Your Step – Session 2

1. Add up a player's scores at the end of a game.
2. Raj is on rock 2. What must he roll to reach 9?
3. Lata is on rock 4. Which die outcomes make her fall off the rocks?

Dicey Dice – Session 1

1. Given a board with placed bids and a dice roll, which player won the bid?
2. Calculate the money won on a winning bid.
3. Calculate total money after one bid.

Dicey Dice – Session 2

1. Money won after a dice roll and bids placed.
2. Which sum is most likely on two dice?
3. In how many ways can you get 11 on two dice?

Table 3. Worksheet questions used across the four-session probability module

Session	Worksheet Questions
Watch Your Step – 1	1. Lata rolls 3 followed by 4. Which rock does she reach?
	2. Raj is on rock 4. Which die outcomes would score one point if he rolls again?
	3. At the end of the round, who receives the bonus and what is their final score?
	4. Lata moves from rock 2 to rock 7 in one roll. What number was rolled?
	5. Raj reaches rock 5 in two rolls. If the first roll is smaller than the second, what could the two rolls be?
	6. If you were on rock 7, would you roll again or stop?
Watch Your Step – 2	1. Lata is on rock 7. Which die outcomes would make her fall into the water if she rolls again?
	2. When Lata is on rock 7, are there more scoring outcomes or falling outcomes?
	3. Raj reaches a total of 6 in two rolls. If the first roll is greater than the second, what could the two rolls be?
	4. Calculate Raj's total score from the given score table.
	5. Calculate Lata's total score from the given score table.
	6. Suppose the die is relabelled with faces $3, 3, 4, 4, 5, 5$. Can a player score on the first roll?
	7. What is the minimum number of rolls needed to reach rock 9?
	8. If Lata is on rock 5, which die outcomes allow her to score?
Dicey Dice – 1	1. Raj places three bets. How much money remains before the dice are rolled?
	2. Given the outcome of the dice, how much money does Raj win?
	3. How much money does Raj have at the end of the round?
	4. Given Lata's bets and the dice outcome, how much money does she win?
	5. Which betting space offers the highest payout?
	6. In how many ways can the sum 12 occur?
	7. If one die shows 1, what values can appear on the other die?
Dicey Dice – 2	1. In how many ways can the sum 5 occur?
	2. In how many ways can the sum 8 occur?
	3. Which sum of two dice is most likely?
	4. Lata bets on both Odd and Even. What is the net outcome of this strategy?
	5. Is the statement "Even sums are more likely than odd sums" correct? Justify your answer.
	6. If the betting space 7 is replaced by Prime, is winning on Prime more likely than winning on 7?

A.2 Worksheet Questions

Worksheets administered after each session include the poll questions as well as some additional questions. The full set of these questions is given in Table 3.

B Student Artefacts

Submissions of homework and challenge questions shared on WhatsApp are shown in Fig. 6.

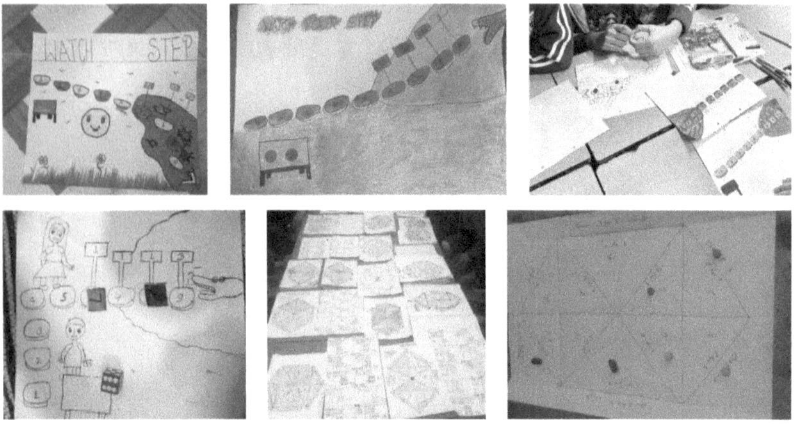

(a) Student-designed Game Boards

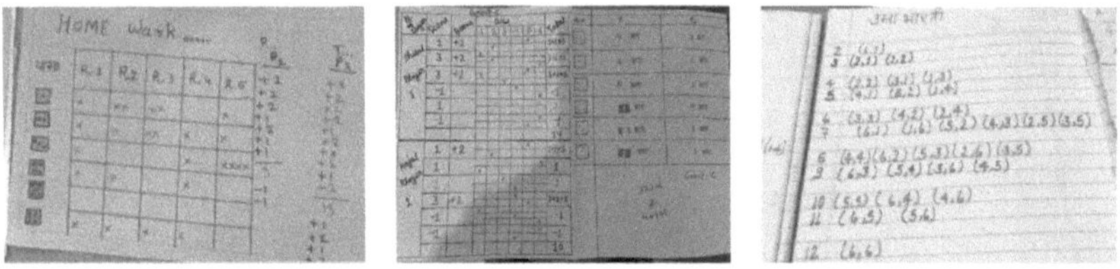

(b) Homework Submissions from WhatsApp Group

(c) Video Screenshots of Bhatiya(l) and Loni(r) Explaining their Design

Fig. 6. Examples of student work collected across the four sessions

References

1. Lecoutre, M.-P.: Cognitive models and problem spaces in "purely random" situations. Educ. Stud. Math. **23**(6), 557–568 (1992)
2. Pratt, D.: Making sense of the total of two dice. J. Res. Math. Educ. **31**(5), 602–625 (2000)
3. Batanero, C., Godino, J.D., Navarro-Pelayo, V.: Combinatorial reasoning and its assessment. In: Gal, I., Garfield, J.B. (eds.) The Assessment Challenge in Statistics Education, pp. 239–252. IOS Press (1997)
4. Lockwood, E., Gibson, B.R.: Combinatorial tasks and outcome listing: examining productive listing among undergraduate students. Educ. Stud. Math. **91**(2), 247–270 (2016)
5. Watson, J.M., Kelly, B.A.: Development of student understanding of outcomes involving two or more dice. Int. J. Sci. Math. Educ. **7**(1), 25–54 (2009)
6. Fischbein, E.: The Intuitive Sources of Probabilistic Thinking in Children. Synthese Library, vol. 85. D. Reidel Publishing Company, Dordrecht (1975)
7. Watson, J.M., Moritz, J.B.: Fairness of dice: a longitudinal study of students' beliefs and strategies for making judgments. J. Res. Math. Educ. **34**(4), 270–304 (2003)
8. Konold, C.: Informal conceptions of probability. Cogn. Instr. **6**(1), 59–98 (1989)
9. Malaspina, M., Malaspina, U.: Game invention as means to stimulate probabilistic thinking. Stat. Educ. Res. J. **19**(1), 57–72 (2020)
10. Batanero, C., Navarro-Pelayo, V., Godino, J.D.: Effect of the implicit combinatorial model on combinatorial reasoning in secondary school pupils. Educ. Stud. Math. **32**(2), 181–199 (1997)

From Cards to Conversations: Tracing Participation Through a Mathematics Game Introduced Online

Adithi Iyer(✉) and Jyothi Krishnan

Indian Institute of Technology Gandhinagar, Gandhinagar, Gujarat, India
{adithi.i,jyothi.k}@iitgn.ac.in

Abstract. This case study examines a screen-mediated introduction to *No Thanks!*, a game in which cards and tokens affect a player's score in opposing ways, so that a costly card can become worth taking as tokens accumulate on it. The game was introduced online to girls attending seven hostels for disadvantaged communities in rural India, who followed the session on a shared television and made cards and tokens from locally available materials. Although connectivity disruptions and the absence of visible follow-up initially suggested that the activity had not continued beyond the session, a later classroom observation and interviews with 17 students showed that they had replayed the game and taught it to peers. They recalled strategies and reasoned about how cards, tokens, and consecutive sequences affected scores; one student described these relationships in terms of positive and negative numbers, although integers were barely mentioned during the session. The case shows how participation that was difficult to detect during a disrupted online session became visible afterward through voluntary replay, peer teaching, and continued mathematical reasoning. It also illustrates why evaluations confined to the instructional encounter may miss later, self-initiated forms of engagement.

Keywords: Situational interest · Mathematics education · Online learning · Game-based learning · Resource-limited settings · Peer participation

1 When the Game Left the Screen

It was a free period in a Grade 7 classroom. The Hindi teacher was on leave, no substitute had been assigned, and students had been instructed to remain in their classroom and "behave properly". Inside the classroom, 35–40 girls were sitting in six separate rows and about 12 boys in two rows. Most of the girls lived in the nearby Kasturba Gandhi Balika Vidyalaya (KGBV). Three groups of girls were seated among the rows with their handmade

M. Löffler et al. (Eds.): CMSC 2026, LNCS 16951, pp. 178–192, 2027.
https://doi.org/10.1007/978-3-032-38887-2_14

copies of *No Thanks!*, a card-and-token game that they had learned to play the previous day. The girls were playing the game and discussing possible moves with one another. Across the room, a group of boys occasionally looked up from their books to watch them. When asked what the girls were doing, the girls interrupted to explain, "We are playing the game you (the researcher) taught yesterday. Look at the cards we made and these small pebbles". They pointed to four other girls and added, *"Didi*[1]*, ye gaaon ke dost hai hamare, hamne inko sikhaya"* ("Didi, these are our friends from the village; we taught them the game".).

When asked whether they had also taught the boys, the girls shyly replied, *"Nahi didi, ham ladko se baat nahi karte"* ("No didi, we do not talk to boys."). The boys offered a similar explanation for not learning the game from the girls: *"Ham ladkiyon se baat nahi karte"* ("We do not talk to girls."). However when invited to learn the game themselves, they immediately nodded.

After lunch, the boys stopped playing in the field and gathered around to learn *No Thanks!* from the researcher. Their sweaty faces showed a mixture of anticipation and shyness. Sitting in a circle, they quickly answered questions such as "Who will pick the next card?", "What will you do if you do not want the card?", and "Who would win the game?" as they learned the rules. Although they actively participated in deciding whether to take or pass a card, they spoke very little to one another during the instruction.

Once the demonstration by the researcher ended, they began playing independently. By then, a few girls had entered the classroom and noticed the boys learning the game. *"Ye toh humein aata hai. Kal online class mein seekha tha"* ("We already know this. We learned it in yesterday's online class"), they remarked. As the game progressed, the usual classroom rows dissolved. The boys sat together playing, while other boys and the girls gathered around to watch.

> *"Aree ye card le lena chahiye tha"* (Oh, you should have taken that card!),
> *"Paas kar do, bada hai card"* (Pass it on, it is a large card!),
> *"Ek card mein itne kankad!"* (So many pebbles on a single card!),
> *"Haan, main le leta hoon ab"* (Yes, I will take the card now).

Such comments soon filled the classroom. The girls and boys, who had earlier stated that they did not talk to one another, were now discussing moves in the game together (Fig. 1. *Adithi Iyer, Field notes, 13 February 2026, Mahuakheda Govt. School, Hoshangabad district, Madhya Pradesh.*)

Children playing a game during a free period is not, in itself, unusual. What stood out was that *No Thanks!* had been introduced only the previous day through an online mathematics session, and no facilitator or researcher had prompted its return. At Mahuakheda[2], the session itself had been difficult to interpret: poor connectivity repeatedly interrupted participation, while in-session polls and photographs shared afterward provided little evidence that the activity had continued. The contrast between what was visible through the online channel and what emerged during the later visit motivates this paper.

(a) Boys at play. (b) Girls prompt and discuss moves.

Fig. 1. Researcher's observation on a school visit.

2 Background and Framing

This paper examines student engagement in a game involving positive and negative quantities. Integers mark a shift towards abstraction in school mathematics and are a well-documented source of difficulty [5,6]. While research has largely examined these difficulties in conceptual terms, for students who may have come to see mathematics as a subject to be completed rather than explored, the first challenge may be less conceptual understanding than voluntary participation. As Schoenfeld observes, alongside the formal curriculum is a "hidden curriculum" that shapes students' beliefs about what mathematics is, how it is learned and done, and who is capable of doing it [17]. From this perspective, interest is not separate from learning; it helps make mathematical engagement possible.

Situational interest provides a useful lens for considering why an activity may capture attention and remain worth pursuing beyond the initial encounter. Features such as novelty, challenge, social interaction, autonomy, and opportunities for discovery can support engagement with an activity [1,2,7,8]. Research on situational interest has largely focused on face-to-face interventions [4,8,11,13], where interaction and physical materials can readily support participation. A growing body of work considers technology-mediated settings including e-learning courses, multimedia environments, digital mathematics games and online video instruction [3,9,14]. These studies, however, largely concern learners

[2] The school referenced in the vignette.

with individual access to devices and content. Less is known about how interest is triggered and sustained in collectively mediated online environments, where groups of learners participate around a shared screen and engagement extends beyond the learner-content relationship. Approaches that focus on engagement observed or reported during the activity may overlook interest expressed later through collective replay, peer teaching, and continued discussion. Using situational interest as a starting lens, we examine two questions: 1) What forms of participation and mathematical reasoning became visible after the screen-mediated session?; 2) What aspects of the activity were prominent in students' accounts of replaying and sharing the game?

3 Setting and Intervention

This lesson formed part of a broader study involving nine sessions on number systems, delivered twice weekly to approximately 600 girls attending Kasturba Gandhi Balika Vidyalayas in rural Madhya Pradesh. Established by the Government of India, KGBVs are residential schools located in Educationally Backward Blocks, identified using indicators such as low female literacy and high gender disparity in literacy. Each hostel accommodates up to 100 students aged 11 to 15 across grades 6 to 8. The students belong predominantly to Scheduled Castes, Scheduled Tribes, Other Backward Classes, minority communities, and other disadvantaged groups [10,12,16]. Seven KGBVs located in the central Madhya Pradesh districts of Sehore, Hoshangabad, and Vidisha were selected for the study. Our sessions were conducted in Hindi and the sites were chosen in part to reduce variation in the language of instruction and interaction; students across these KGBVs could participate in a common form of Hindi. Although the participants resided in KGBV hostels, they attended nearby government secondary schools that also enrolled boys and girls from surrounding villages. The opening episode described in this paper was observed in one such classroom.

(a) (b)

Fig. 2. (a) The session arrangement planned with wardens and teachers. (b) The setup in one of seven KGBVs, as shared by the warden.

After school hours twice a week, students at each hostel gathered around a shared LED television in their respective common rooms to attend these sessions.

An exploratory study conducted before the intervention had identified that these televisions supported YouTube and audio playback but lacked cameras, microphones, personal logins, and app installation capabilities. Consequently, a set-up was planned and communicated to wardens (Fig. 2). Interaction with the facilitator was enabled through a smartphone connected to Zoom and placed near the television, using either Wi-Fi or a mobile hotspot depending on local availability. Before each session, wardens communicated the materials required, all of which could be sourced locally. The activities were designed around novelty, hands-on making, challenge, and collaboration. Students constructed materials, played and discussed the games in groups, and responded to polls using cards labelled A-D. Worksheets were provided after each session (Appendix B).

4 Data and Ethical Considerations

This case study focuses on one of the nine online sessions, delivered across seven KGBVs, with detailed follow-up evidence from two sites, drawing on the session recording, field notes from the Mahuakheda visit, and interviews (Appendix A for the protocol). Seventeen students were interviewed, selected by wardens to represent a range of school performance: eight in person at Mahuakheda and nine online at Barkheda Hasan. We then reviewed these sources for evidence of participation during the session, replay after the session, peer teaching, retention of materials, strategic reasoning, and connections to school mathematics. We compared what had been visible through the online channel with what was later observed or reported. The data were reviewed to identify characteristics of students' interactions during the sessions, the affordances that emerged through these interactions, the conditions under which these affordances arose, instances of voluntary replay, peer teaching, strategic reasoning, and connections to mathematical ideas. We compared engagement visible during the online session with engagement observed or reported afterward.

Ethical approval was obtained from the IIT Gandhinagar Institute Ethics Committee and the State Education Department of Madhya Pradesh. Written assent was obtained from students and consent from their wardens. All names are pseudonyms, photographs are reproduced only with permission, and the data were securely stored and used solely for this study.

5 The Game and Its Mathematics

Games have been described not merely as instructional tools but as socially organised environments involving interaction, coordination, decision-making, and participation [15]. Games with rules require players to negotiate turns, anticipate outcomes, and respond to peers, creating opportunities for collaboration and discussion. For this session, we selected *No Thanks!*, designed by Thorsten Gimmler and used with his permission, because it combines social play, strategic decision-making, and a mathematical structure that can be experienced without formal explanation.

(a) A five-player game state.

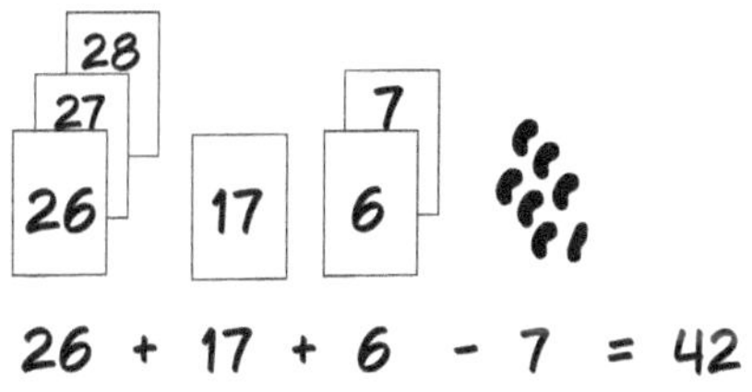

(b) An illustrative scoring rule.

Fig. 3. Illustrations from the *No Thanks!* worksheet.

The game uses numbered cards from 3 to 35 and a supply of tokens. We played in groups of six, with nine tokens per player. Nine cards are removed at random before play begins, so part of the deck remains unseen. On each turn, the top card is revealed, and the active player either accepts it together with any tokens accumulated on it or passes by placing one of their own tokens on the card. Figure 3a shows an example game position. The card continues around the group until someone accepts it, after which a new card is revealed. At the end, players add the values of the cards they hold, counting a consecutive run only by its lowest card, and subtract one point for each token they retain. The lowest score wins (Fig. 3b).

The game creates opportunities to experience several mathematical ideas through play. Cards and tokens move a score in opposing directions: accepting a card raises it, while retaining tokens lowers it. These effects can be represented using positive and negative quantities, although this connection was discussed only briefly during the session. What makes the game more than an exercise in arithmetic is that these quantities interact. Passing costs a token immediately but may avoid a larger cost; as tokens accumulate, a previously undesirable card may become worth accepting. A high card may also be worth taking if it completes a consecutive run, since only the lowest card in the run is counted, but the required card may be among the nine removed from the deck. Few decisions have a single correct answer. Players must coordinate card values, accumulated tokens, possible runs, unseen cards, and their limited supply of tokens.

The game also provides a bridge between informal gameplay and formal mathematics. Situations arising during play can be revisited through questions about strategy, scoring, and representation. In follow-up worksheets (see B), students predict how accepting a card and gaining or spending tokens changes a player's score, and represent these changes using expressions such as $(20+6-4)$, where the cards contribute positive values and the tokens reduce the total. Gameplay can therefore become a starting point for discussions of integer operations, optimisation, and strategic reasoning rather than an activity separate from mathematics. The mathematics of *No Thanks!* emerges through the decisions players repeatedly make. In what follows, we examine how students engaged with these relationships during the session and afterward.

6 Engagement Visible During the Session

The activity began with students constructing the game materials using paper cards and locally available objects such as pebbles and pulses as tokens (Fig. 4a). As they prepared the materials, groups discussed the task with one another and several students moved closer to the camera to display what they had made (Fig. 4b). During the explanation of the rules, students remained attentive while simultaneously handling their materials. Across schools, students responded to comprehension checks with thumbs-up gestures (Fig. 4c), attempted each step of the demonstration as it was introduced, and increasingly rehearsed the moves being shown on the screen. They shuffled their cards, arranged them into piles, and imitated gameplay even before the demonstration had concluded. At one KGBV, a student reached forward to perform a demonstrated move before a peer reminded her to continue watching the explanation, suggesting that students were already anticipating playing the game themselves.

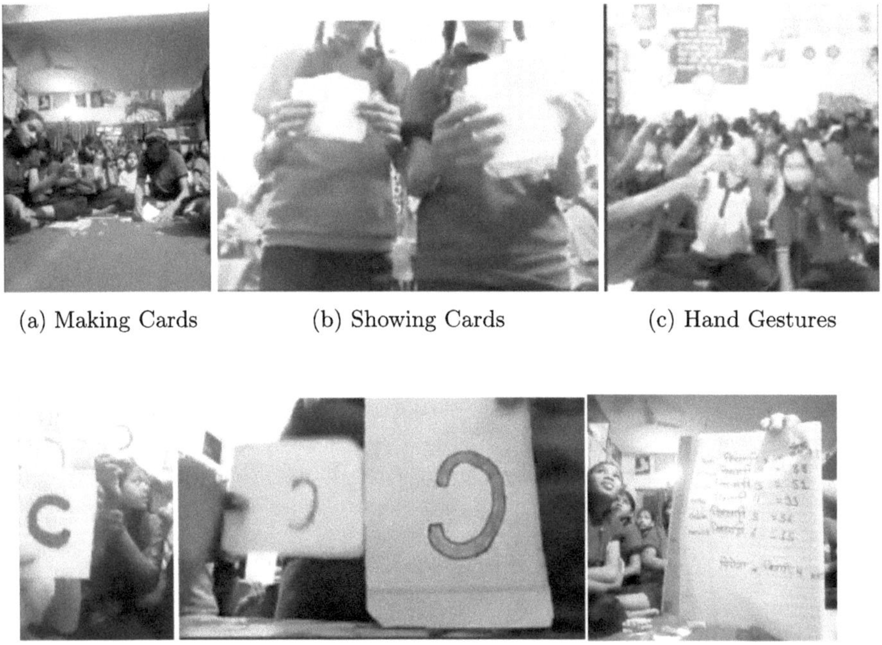

(a) Making Cards (b) Showing Cards (c) Hand Gestures

(d) Option Cards (e) Sharing scores

Fig. 4. Students participating during the online *No Thanks!* session.

Although network interruptions occurred at several schools, students and staff repeatedly attempted to reconnect and continue the session. Students also held their score cards up to the camera to share their progress with the researcher (Fig. 4e). Where connections were restored, groups resumed participation from where they had stopped, and interviews later confirmed that students remembered these disruptions. Nevertheless, neither the observations nor the interviews suggested that the interruptions discouraged participation during the session.

In contrast to the other sessions in the series, this session was not immediately followed by the photographs and videos of gameplay that wardens usually shared. *No Thanks!* was the seventh of the nine sessions, and by that point a routine had developed in which staff sent images and short clips of students playing and explaining the activity soon afterward. Their absence here, combined with the connectivity disruptions during the session, initially suggested the activity generated little engagement.

7 Engagement Beyond the Session

Contrary to the lack of post-session evidence, interview accounts suggested that students had enjoyed the activity. Kriti (Grade 8) recalled, *"Everyone was playing, everyone was enjoying it, and everyone was laughing."*. When asked whether the laughter was unrelated to the game, she explained that students were chanting *"No Thanks, even I don't want this card! No Thanks!"* and joking about how long a card might continue circulating before someone finally accepted it because of the tokens accumulated on it.

Students prepared their own cards and tokens from locally available resources: notebook or chart paper for the numbered cards, and pebbles or pulses such as *rajma* for the tokens. The materials were referred to repeatedly during interviews and, in several cases, physically displayed by students as they explained the game.

At Mahuakheda, one reason for the quiet immediately afterward became clear only in interviews: the session had coincided with the assistant warden's birthday, and the girls spent the evening preparing decorations, making gifts, and looking forward to a special meal. This was a site-specific circumstance rather than a general explanation, but it illustrates a broader point, competing social priorities, invisible through the online channel, could displace an activity without diminishing interest in it. Interviews at both sites indicated that students had enjoyed the game and replayed it: during the school visits the students described replaying it during free periods and teaching it to peers who had never attended the session, the episode with which this report opens. Consistent with these accounts, thirteen of the seventeen students interviewed reported playing again the following day. Five students specifically mentioned playing the game with friends who had not participated in the original session. The activity had not ended with the disrupted online session; it had continued and spread through the students' own social world.

7.1 Strategic and Mathematical Reasoning

Students did not remember the activity in a uniform way. Some recalled making the cards and collecting pebbles; others began by describing the numbered cards, the setup, or the turn structure; still others moved quickly to strategy. This indicates that engagement did not have a single form. The activity offered multiple entry points into participation, and only some of those entry points were explicitly mathematical at first.

Interview accounts suggested that students had developed an understanding of the game's scoring structure. Fourteen students were able to explain how taking or giving away a card or token affected a player's score, and one student explicitly related these changes to the positive and negative numbers studied in school mathematics.

Asked to explain the game, most students began with the setup, forming groups, distributing tokens, shuffling the cards, and removing nine cards, before describing the sequence of play. Several, however, went beyond recalling the rules to discuss strategies they had used while playing. Rohini (Grade 7), for example, explained that the first player could either take a card numbered twenty-five or say "No Thanks" and pass it on. To the question of whether she would take twenty-five, she reasoned, *"No, it is a large number. The player with the smallest total wins, so we would not take it."* Rather than treating twenty-five as a fixed part of the rules, she evaluated it against the winning condition. Similar reasoning appeared in other interviews, although what counted as 'large' or 'small' varied across students. Aarti (Grade 8) remarked, *"We thought we should take small numbers, like 5 or 10."*. The coordination of card values, tokens, and consecutive runs was also visible in this exchange with Priyam (Grade 7):

> Interviewer: *Usko 22 rakhna chahiye kya?* (Should she keep the 22?)
> Priyam: *Ji ma'am.* (Yes ma'am.)
> Interviewer: *Kyu?* (Why?)
> Priyam: *23 rakha hai toh 22 bhi rakh sakte hai.* (She already has 23, so she can keep 22 too.)
> Interviewer: *Kyu?* (Why?)
> Priyam: *Ma'am woh jyada badi sankhya nahi hai, par kankad kam hai, aur aakri mein 23 ko nahi ginenge.* (It isn't a very large number, there are few pebbles on it, and at the end 23 won't be counted separately.)

A number in the twenties was generally treated as undesirable and "large," yet students were willing to accept larger numbers depending on the tokens accumulated and the consecutive-number rule. They were, in other words, beginning to coordinate card values, tokens, and the run rule simultaneously, though with differing sophistication.

Students reasoned, when prompted, about how cards and tokens move a score in opposite directions. Neha (Grade 6) initially thought tokens were simply subtracted at the end, but when presented with a sequence of hypotheticals, worked out that receiving a token lowers the score while giving one away raises it: *"When I take* rajma *(kidney beans used as tokens) the score goes down, when*

I give it away it goes up" and contrasted this with cards: "*Keep a card and it goes up; give it away and it goes down*", tracing how the score changed as objects moved between players, rather than reciting a rule. Kriti went further, mapping the game onto school mathematics:

> Interviewer: *Toh hum jo karte hai woh tumko lagta hai ki ganit hai?* (Do you think what we're doing is mathematics?)
> Kriti: *Haan.* (Yes.)
> Interviewer: *Kaise?* (How?)
> Kriti: *Ismein bhi jaise humne dekha ki rajma lete hai toh ghata hai aur dete hai toh badhta hai, toh jaise rinatmak dhanatmak sankhya rehta hai, yeh bhi waisa hi hai.* (Here too, taking tokens lowers it and giving them away raises it, it's like positive and negative numbers.)
> Interviewer: *Ismein minus vaali konsi sankhya thi?* (Which quantity here was the negative one?)
> Kriti: *Rinatmak rajma the, aur dhanatmak cards the.* (The tokens were the negative quantities, and the cards the positive ones.)

Kriti identified tokens with negative and cards with positive quantities and reasoned that the score changed according to whether each was added or removed, connecting the game's mechanics to the formal idea of signed numbers she had met in school. At Mahuakheda the worksheet had not yet been shared by the warden, so this connection, for Kriti at least, was drawn from the game itself rather than prompted by the worksheet's explicit framing of tokens as -1. Other students demonstrated strategic coordination of card values, tokens, and consecutive runs without explicitly naming the underlying mathematics.

7.2 A Social Boundary Temporarily Crossed

At Mahuakheda, the game also drew in students who had not attended the online session. Girls had taught village friends, while the researcher later introduced the game to boys in the same classroom. Once play began, girls watching from nearby commented on the boys' decisions. This was a single, researcher-mediated episode and does not indicate a lasting change in gender relations. Within the case, however, it shows how the game created a temporary shared focus among students who had initially described themselves as socially separate. This episode resonates with Thorne's account of gender "borders" in children's social worlds, in which children organise into same-gender groups and maintain distinctions through everyday interaction [18].

8 Ownership, Circulation, and Portability

Taken together, these observations suggest three features that may have contributed to the activity's persistence beyond the session, none of which our online indicators could readily capture. The first is *ownership through making*. Because students built the cards and gathered the tokens themselves from locally available materials, the materials appeared to become closely tied to their experience of the game; students referred to them repeatedly in interviews and often displayed them while speaking (Fig. 5). The second is peer circulation. The game spread laterally from girls who had attended the session to village friends through peer teaching. The later researcher-mediated introduction to the boys extended the activity to another group and created a temporary shared focus in the classroom. The third is *portability*. An unplugged game made of paper and pebbles could be replayed in a free period with no screen, no connectivity, and no facilitator, precisely the conditions under which it resurfaced. In this light, the persistence observed during the session, such as repeated attempts to reconnect after network failures, can be interpreted as early signs that participants considered the activity worth continuing.

Fig. 5. Students display their handmade copies of the game during interviews.

Viewed through the lens of situational interest, the case draws attention to forms of engagement that appeared only after the facilitated encounter. The relevant features were not confined to what happened on screen but continued later. Interest, in this case, was maintained not simply through attention to the screen but through the social and material afterlife of the game: the cards could be kept, the rules could be taught, and the activity could be replayed without the original facilitator.

The case also offers practical lessons for collectively mediated online mathematics activities. First, activities intended to outlast a session should be designed for detachment from the screen and built around materials that learners can create and recreate themselves. Second, re-engagement may be delayed rather than immediate; competing social priorities can temporarily displace an activity without diminishing interest in it. Finally, researcher-visible indicators such as photographs, messages, and immediate replay may provide only a partial picture of participation. Understanding what happens beyond the session may require forms of follow-up that extend beyond the communication channel itself.

9 Limitations and Conclusion

This paper focuses on a single session with a relatively small, warden-identified interview sample at two of the seven sites; the cross-gender interaction rests on one observed episode; and, without a comparison condition, we cannot separate the effects of the game's mathematical content, social form, and novelty. We also do not know how long the game continued to circulate after the observed follow-up. Even with these limitations, the case suggests that the significance of a screen-mediated mathematics activity may not be visible through the screen itself. The contribution of this case is not simply that students enjoyed a mathematical game, but that their interest became visible beyond the instructional encounter through voluntary replay, peer teaching, and continued reasoning about the game's mathematical structure. Evaluations confined to the session itself may miss these later and collectively expressed forms of mathematical engagement. Future work could examine whether similar forms of replay, peer circulation, and mathematical reasoning emerge across other activities and sites, and how long they persist beyond the facilitated session.

Acknowledgements. The authors are grateful to Prof. Malay Dhamelia for his valuable comments and suggestions. We thank the Centre for Creative Learning, IIT Gandhinagar, for providing an environment that supported the design and implementation of this work. The first author thanks her advisers, Prof. Sameer Sahasrabudhe and Prof. Manish Jain, for their guidance throughout the study.

A Semi-structured Interview Protocol

The following guide structured the interviews with the 17 students. It was used flexibly: questions were reordered, rephrased in Hindi, or skipped where a student had already addressed them, and follow-up probes were added to pursue students' own reasoning. Interviews were conducted in Hindi; the English here is a working translation.

Opening and Recollection.

1. Do you remember the online class we did with the cards and tokens? What did we do that day?
2. What did you use to make the cards and tokens? Where did you get the materials?
3. Who did you play with? How many of you were in a group?

Rules and Setup.

4. Can you explain the game to someone who has never played it?
5. How does the game begin? How many cards are removed, and how many tokens does each player get?
6. It is your turn and you draw a card. What are your choices?
7. How does the game end? How do you decide who has won?

Strategy and Reasoning.

8. Suppose the card on top is 25. Would you take it or pass it? Why?
9. Does it ever make sense to take a large-numbered card? When?
10. What do the tokens on a card change about your decision?
11. You already hold the card 23 and the 22 appears. What would you do? Why?

Connections to Mathematics.

12. Do you think what we did in this game is mathematics? How?
13. When you take a token, what happens to your score? When you give one away?
14. When you keep a card, what happens to your score? When you pass it on?
15. Have you seen anything like this—numbers that go up and down—in your school mathematics?

Replay and Circulation.

16. Did you play the game again after the online class? When, and with whom?
17. Did you teach the game to anyone who was not in the class?
18. Did you keep the cards and tokens you made? Can you show them to me?

B Sample Worksheet Questions

The following questions are representative examples from the worksheet used during the *No Thanks!* intervention. Figures like in Fig. 3 were included for students to analyse. Note that we use buttons as tokens with values -1.

Representing Scores

1. At the beginning of the game, each player has 11 buttons. Write your starting score as a number.
2. The first card is 4. You decide to take it. What is your new score? How do you know? Write an equation showing your new score.
3. The next card is 34. You decide not to take it and instead give away one button. What is your new score? Write an equation representing your score.

Addition with Cards and Buttons

4. Suppose you have card 34 and you take a card numbered 3 and the total score is 37 points. If we write the action of taking as "+", then this can be written as:

$$\text{card } 34 + \text{card } 3 = \underline{37} \text{ points}$$

Complete the following using this:

$$\text{card } 28 + \text{card } 5 = \text{\_\_\_\_\_\_ points}$$
$$\text{card } 28 + 3 \text{ buttons} = \text{\_\_\_\_\_\_ points}$$
$$\text{card } 20 + \text{card } 6 + 4 \text{ buttons} = \text{\_\_\_\_\_\_ points}$$

Subtraction with Cards and Buttons

5. Suppose your score is 37 points, made from card 34 and card 3. If you give away card 3,the total score will be 34 points. If we write the action of giving away as "-", then this can be written as:

$$37 \text{ points} - \text{card } 3 = \underline{34} \text{ points}$$

Complete the following using this:

$$28 \text{ points} - \text{card } 5 = \text{\_\_\_\_\_\_ points}$$
$$42 \text{ points} - 3 \text{ buttons} = \text{\_\_\_\_\_\_ points}$$

Conceptual Understanding

6. When you take a card, your score:

(i) increases (ii) decreases (iii) does not change

7. When you give away a card, your score:

(i) increases (ii) decreases (iii) does not change

8. When you take a button, your score:

(i) increases (ii) decreases (iii) does not change

9. When you give away a button, your score:

(i) increases (ii) decreases (iii) does not change

References

1. Ainley, M.: Students' interest and engagement in classroom activities. In: Handbook of Research on Student Engagement, pp. 283–302. Springer (2012)
2. Dewey, J.: Interest and Effort in Education. Houghton Mifflin, Boston (1913)
3. Dousay, T.A., Trujillo, N.P.: An examination of gender and situational interest in multimedia learning environments. Educ. Tech. Res. Dev. **67**(2), 307–327 (2019)
4. Flowerday, T., Schraw, G.: Teacher beliefs about instructional choice: a phenomenological study. J. Educ. Psychol. **92**(4), 634–645 (2000)
5. Gallardo, A.: Negative numbers in the teaching of arithmetic. Repercussions in elementary algebra (1995)
6. Gallardo, A.: The extension of the natural-number domain to the integers in the transition from arithmetic to algebra. Educ. Stud. Math. **49**(2), 171–192 (2002)
7. Hidi, S., Baird, W.: Interestingness - a neglected variable in discourse processing. Cogn. Sci. **10**(2), 179–194 (1986)
8. Hidi, S., Renninger, K.A.: The four-phase model of interest development. Educ. Psychol. **41**(2), 111–127 (2006)
9. Lange, C.: Examining situational interest and its relationship with self-efficacy in asynchronous and synchronous video lectures. Int. J. Educ. Technol. High. Educ. **21**(1), 1 (2024)
10. Ministry of Education, Government of India: Ministry of education to provide ICT labs and smart classrooms in all functional kasturba gandhi balika vidyalayas and hostels (2024). https://www.pib.gov.in/PressReleasePage.aspx?PRID=2030430. Press Information Bureau, Accessed 1 June 2026
11. Mitchell, M.: Situational interest: its multifaceted structure in the secondary school mathematics classroom. J. Educ. Psychol. **85**(3), 424 (1993)
12. National Institute of Educational Planning and Administration: Kasturba gandhi balika vidyalaya (kgbv) (2024). https://www.niepa.ac.in/KGBV/index.html
13. Renninger, K.A., Bachrach, J.E., Hidi, S.E.: Triggering and maintaining interest in early phases of interest development. Learn. Cult. Soc. Interact. **23**, 100260 (2019)
14. Rodríguez-Aflecht, G., Ramani, G.B., Siegler, R.S.: Continuity and change in fifth graders' mathematics interest and knowledge in a six-week digital game intervention. J. Educ. Psychol. **110**(8), 1171–1188 (2018)
15. Rye, S., Sousa, M., Sousa, C.: Introduction to game-based learning. In: Transformative Learning Through Play: Analogue Games as Vehicles for Educational Innovation, pp. 29–68. Springer (2025)
16. Samagra Shiksha Gujarat: Kasturba gandhi balika vidyalaya (kgbv) (2024). https://samagrashiksha.ssagujarat.org/en/kgbv. Accessed 1 June 2026
17. Schoenfeld, A.H.: Learning to think mathematically: problem solving, metacognition, and sense making in mathematics (reprint). J. Educ. **196**(2), 1–38 (2016)
18. Thorne, B.: Gender play: girls and boys in school. In: Gender Play. Rutgers University Press (2024)

Technology-Enhanced Tools

Integrating Immersive Technologies Into Secondary Mathematics: from AR Applications to Teachers' Training

Valeria Andriano(✉), Monica Barbero, Matteo Bramardi, and Walter Dambrosio

Università di Torino, Turin, Italy
{valeria.andriano,monica.barbero, matteo.bramardi,walter.dambrosio}@unito.it

Abstract. This paper presents the design and initial implementation of a professional development (PD) program aimed at supporting secondary school mathematics teachers in the meaningful integration of augmented reality (AR) technologies. Building on research on technology integration in mathematics education, the program addressed the growing need for teacher preparation in the use of immersive technologies for STEM education. The training focused on two AR applications conceived within the University of Turin: xTimeTracer, designed to explore motion and its graphical representations through embodied interaction, and GraphSurfing, which supports an embodied approach to the concept of derivative. The PD is situated within a broader project involving the design, implementation and evaluation of classroom activities. Drawing on this experience, the training program focused on ensuring technical autonomy, providing a theoretical foundation, and guiding teachers through the collaborative design and revision of their own classroom materials. The qualitative analysis of a concluding questionnaire, tracking actual classroom implementation and teacher perceptions, and of tutors' field notes provides the first results of the project. Overall this contribution presents the two AR applications and highlights the importance of practice-oriented and theoretically grounded PD initiatives for the sustainable integration of AR technologies in mathematics education.

Keywords: Augmented reality · Teaching with technologies · Teachers' professional development · Embodiment · Semiotic registers

1 Introduction and Literature Review

The integration of digital technologies into mathematics education has long been recognised as a powerful means of enhancing teaching and learning. Over the past decades, research has demonstrated that technology can support students' conceptual understanding, foster mathematical exploration, and enhance interactive learning environments [16]. More recently, emerging technologies such as

M. Löffler et al. (Eds.): CMSC 2026, LNCS 16951, pp. 195–207, 2027.
https://doi.org/10.1007/978-3-032-38887-2_15

Augmented Reality (AR) and Virtual Reality (VR) have further expanded the possibilities for interaction and visualisation in mathematics education, attracting increasing attention from researchers and educators alike [7,16]. A growing body of literature has highlighted the effectiveness of AR and VR environments in helping students construct meaning around mathematical concepts and engage with abstract ideas in more intuitive and embodied ways [4,5].

Despite these promising developments, enabling teachers to integrate technology purposefully and effectively into mathematics instruction remains a major challenge. The foundational work of Mishra and Koehler [13] introduced the Technological Pedagogical Content Knowledge (TPCK/TPACK) framework to conceptualise the forms of teacher knowledge required for meaningful technology integration. Building on Shulman [15] and Niess [14], TPCK emphasises the complex interplay among technological, pedagogical, and content knowledge [2]. According to Mishra and Koehler [13], effective teaching with technology requires knowledge of how concepts can be understood through technological tools, how pedagogical strategies can leverage these tools constructively, and how technology can address students' learning difficulties while building on their prior knowledge and epistemological understandings.

Within this perspective, the integration of Extended Reality (XR) technologies—including AR and VR—into mathematics education introduces new demands on teachers' professional knowledge and practice [9,11]. Recent studies illustrate both the educational potential and the practical challenges associated with these technologies. For instance, the study by Lo and colleagues [12] reported significant improvements in students' conceptual understanding of locus concepts through XR learning environments, while simultaneously identifying implementation barriers related to accessibility and teacher readiness. Teacher readiness consistently emerges as a critical factor in the successful adoption of educational innovations. Research in mathematics education increasingly highlights the importance of professional development (PD) programs that equip teachers with both the technical competencies and the pedagogical confidence necessary to integrate advanced technological tools into classroom practice.

Collectively, this body of research underscores the central role of teacher training in supporting effective technology integration. Effective PD initiatives should not only provide technical training but also foster pedagogical integration strategies closely connected to authentic classroom practice. Previous studies [8] suggest that successful PD programs should engage teachers actively in lesson design, hands-on experimentation with technological tools, collaborative reflection, and iterative adaptation of instructional practices. In this perspective, teacher PD is not limited to fostering technical proficiency with XR tools. Rather, it involves supporting teachers in the design of classroom activities in which technological affordances, mathematical meaning-making, and pedagogical choices are developed in close relation to one another.

Against this background, this paper reports the design of an in-service PD program aimed at supporting mathematics and physics teachers in the collaborative design and classroom implementation of XR-based activities grounded in embodied approaches to mathematics learning. The program builds on XR tools already available in schools through national funding initiatives, but still only marginally integrated into everyday teaching practice.

In this contribution, following a brief overview of the employed applications and previously tested classroom activities, we describe the design principles underlying the program and the structure of the PD activities, and we discuss the program's initial results based on teachers' perceptions, as reflected in their responses to a questionnaire. In doing so, we aim to contribute to the growing discussion on how to prepare teachers for the pedagogically meaningful integration of XR technologies in mathematics classrooms.

2 Theoretical Framework

Embodiment. Drawing on Bruner's research and embodied cognition theory [10], Tall [17] argues that mathematical thinking develops from intuitive and perceptual experiences into symbolic manipulation and formal deductive reasoning. His work highlights the importance of embodiment, whereby learners build mathematical meaning from physical actions, visualisations, and mental images before engaging with abstract concepts. In the context of calculus, Tall [17] showed that students often rely on embodied intuitions about motion, change, and continuity, which can support or sometimes conflict with formal mathematical definitions. His framework describes mathematical thinking as developing through three interconnected worlds: the embodied world, where concepts arise from perception and physical experience; the symbolic world, where mathematical symbols and procedures are used to represent and manipulate ideas; and the formal world, where mathematics is organized through definitions, axioms, and logical proofs. According to Tall, successful learning of calculus requires students to move flexibly between these worlds, linking intuitive notions of change and accumulation with symbolic techniques and formal reasoning.

Semiotic Registers. Duval [6] argues that all mathematical activity depends on the use of semiotic representations. Mathematical objects can be expressed through a variety of semiotic representations, including natural language, symbolic notation, and visual forms such as diagrams and graphs. These representations are essential for the development, communication, and understanding of mathematical knowledge, and only systems that permit transformations qualify as semiotic registers.

Within this perspective, Duval [6] distinguishes between two types of transformations: treatments and conversions. Treatments occur within a single register, for example performing algebraic manipulations on an equation, while conversions involve moving between different registers without changing the underlying mathematical object. Such conversions are particularly important because they can make certain mathematical properties more apparent. For example, representing a function graphically may reveal characteristics such as monotonicity or boundedness that are less evident in its algebraic form.

Documentational Genesis. The concept of documentational genesis [8] is a theoretical framework developed within mathematics education research to explain the process through which teachers transform resources into documents through use, adaptation, reflection and classroom experience. In this framework, resources are any material or symbolic artefact available to teachers (such as textbooks, digital tools, lesson plans, XR applications, worksheets, online platforms) whereas documents are the result of a teacher's interaction with these resources combined with their schemes of utilisation (their ways of using, adapting, and integrating the resources into teaching practice).

The theory emphasises two processes: instrumentation and instrumentalization. In instrumentation the resource influences the teacher's activity and thinking. For example, an AR geometry app may encourage a teacher to emphasize spatial visualisation or dynamic manipulation. In instrumentalization the teacher adapts and modifies the resource according to instructional goals and classroom constraints.

In mathematics teacher education, documentational genesis helps explain how teachers develop professional knowledge when integrating technology. Rather than simply "using" technological tools, teachers gradually construct resource systems that reflect their pedagogical beliefs, mathematical goals, and classroom experiences.

3 Applications and Classroom Activities

3.1 Applications and Their Main Features

The applications driving this work, namely xTimeTracer and GraphSurfing, were co-developed by the University of Turin and the Italian startup LevelUP. They are currently available for Oculus 2 as well as Meta Quest 2, 3, and 3 s headsets. The University's involvement in the development process allowed the software to be tailored to specific educational needs, as highlighted below.

XR experiences can be shared within small groups by projecting the headset view onto an external screen. While testing our activities, we found that wired projection via the Meta Quest Developer Hub application delivers a smooth, real-time experience without rapidly draining the device's battery. This setup allowed us to design longer activities while still ensuring that each student adhered to the recommended screen-time limits.

xTimeTracer and AnReAL xTimeTracer immerses users in an AR environment, where they can see and interact with a small sphere attached to a straight line spanning from -1 m to +1 m.

A space-time graph is displayed directly above this line. Using two separate buttons on the headset controllers, users can move the sphere along the line as if it were a real object and track its position, which is plotted in real time on a position-time graph. The application can track other motion characteristics and up to five graphs can be displayed at the same time, including velocity-time

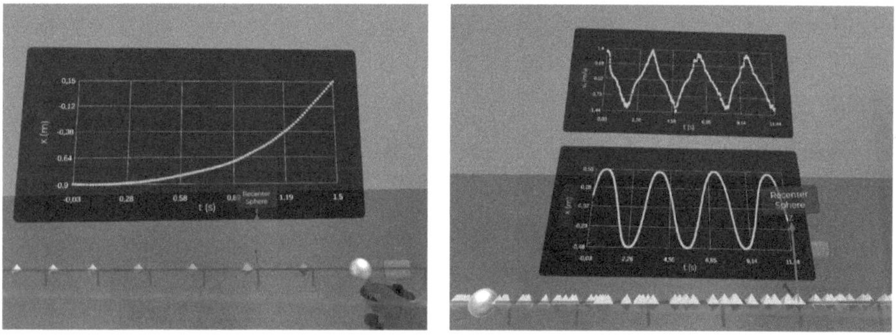

Fig. 1. xTimeTracer: moving the sphere (left), position and velocity-time graphs (right).

and acceleration-time plots. Furthermore, users can set their reference system by repositioning the line (see Fig. 1).

The University of Turin conceived xTimeTracer starting from AnReAL, a preexisting application that employs the headset's built-in sensors to analyse the three-dimensional motion of each headset controller. To provide a suitable environment for secondary school education, the movement was restricted to a single dimension. Consequently, rather than tracking the controllers directly, xTimeTracer features a sphere constrained to a straight line.

GraphSurfing. GraphSurfing immerses users in an AR environment where they can display the graph of a real function by either selecting it from a list of notable examples or typing it in manually.

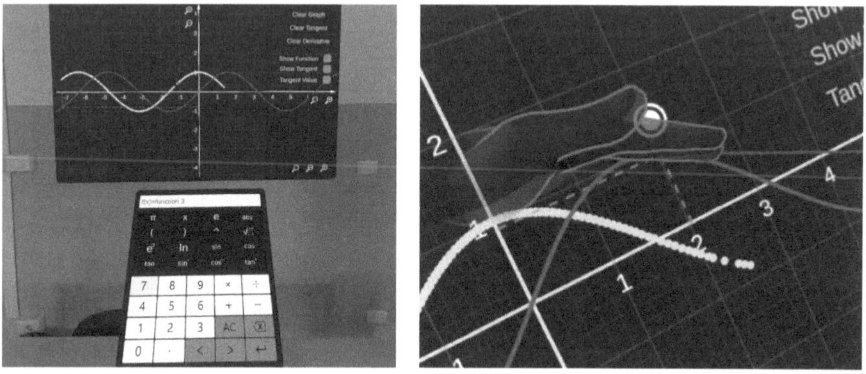

Fig. 2. GraphSurfing: a function and its derivative (left), hand "surfing" a graph (right).

By "surfing" with their hand along the graph, as if it were its tangent at any given point, users can discover the graph of its derivative (see Fig. 2). Users can toggle the visibility of the function's analytical expression, as well as of the tangent line and the slope of its graph at any point. It is also possible to reposition the graph and to zoom in or out.

GraphSurfing does not require the use of physical controllers, as it relies on the headset's built-in hand tracking system. It was designed by the University of Turin building on Tall's seminal work [17] on the embodiment of the derivative.

3.2 Classroom Activities

The teacher PD program is grounded in our experience within CRT Foundation's Progetto Diderot, which focuses on developing and implementing innovative educational activities across various secondary schools in the Piedmont and Aosta Valley regions, Italy. In particular, we used xTimeTracer and GraphSurfing to design a set of activities for upper secondary STEM education.

While each activity targets a specific mathematical or physical concept, they are all designed to maximize their effectiveness across different contexts and they share a common structure and theoretical framework. Based on embodied cognition [17], the design of the activities aims to anchor the understanding of key mathematical concepts in students' own physical experiences. Moreover, following the theory of semiotic registers [6], specific tasks encourage students to translate these concepts back and forth among a variety of representations: their kinesthetic experience is explicitly linked to the graphic and symbolic registers, as well as to natural language as they discuss their findings.

Each activity accommodates up to 30 students and can be adapted to last between 100 and 120 min. The session begins with a brief, whole-class introduction on how to use the headsets, after which the main activity is carried out in small groups of 3 to 5 students. Within each group, students take turns wearing a headset while their view is shared with their peers via screen mirroring. A structured series of tasks guides them to explore and debate with one another before presenting their conclusions to the instructors. Each activity is completed by a brief whole-class discussion.

xTimeTracer activities respectively focus on exploring the space-time graph of one-dimensional motions; understanding the role of velocity in the space-time graph of one-dimensional motions; understanding the core features of the space-time and velocity-time graphs of harmonic motions; exploring the relationship between the space-time and velocity-time graphs of one-dimensional motions. GraphSurfing provides a more targeted embodied experience, allowing students to explore the geometric meaning of the derivative.

4 The Teacher Professional Development Program

Following the initial phase of developing and testing activities with XR headsets in upper secondary schools, the implementation of a dedicated teacher PD program represented the next logical progression.

The primary motivation for this program is to scale up the adoption of STEM educational practices utilising XR headsets within high school mathematics and physics curricula, and to ground these activities in pedagogical frameworks, particularly embodiment.

The PD training program also aims to establish a multi-level network: within individual school STEM departments, between different schools, and between schools and the University. Collaboration between teachers facilitates the integration of these digital tools into daily classroom activities, while at the institutional level, this network serves to promote future projects and involve schools directly in academic research.

Moreover, a key pragmatic motivation for this program stems from the availability of XR headsets in Italian schools. Following the allocation of funds under the National Recovery and Resilience Plan (PNRR), a number of secondary schools acquired XR headsets in an attempt to transform traditional classrooms and laboratories into innovative and immersive learning environments. However, these resources frequently remain underutilised. Informal feedback gathered from conversations with secondary school teachers indicates that when headsets are used, their application is usually confined to simple visual immersion, thereby neglecting the potential of hardware sensors for data collection and body tracking for embodiment. The training program was therefore designed to provide teachers with tools to explore new ways of approaching XR headset usage for active mathematics and physics education.

The training program was delivered as a 15-hour PD course structured across four meetings. Two trainers were assigned per school: one focused on hardware and software training, while the other presented the pedagogical and mathematical foundations, and shared their prior expertise in designing immersive activities..

The PD was structured around four core pillars: enabling teachers on a technical level to ensure hardware and software autonomy; sharing past experience in designing educational activities for XR headsets; allowing teachers to actively design their own activities and test them in the classroom; discussing the outcomes and gathering feedback. Overall, the program aimed to shape teachers' perceptions of XR headsets, helping them transition from viewing the hardware as a technological novelty to actively integrating it as a meaningful resource for mathematical meaning-making.

Technical Autonomy. During the initial sessions, teachers were thoroughly guided in the use of the XR headsets and, subsequently, xTimeTracer and GraphSurfing were introduced. During this phase, teachers also learned how to troubleshoot common problems that might occur during normal classroom hours, ensuring they could manage the equipment independently. Additionally, school technical personnel received training to ensure they could effectively assist teachers with hardware management.

Experience Sharing and Theoretical Framing. The training program did not approach XR technologies as stand-alone technological tools. Rather, the pedagogical framing aimed to support teachers in reflecting on how immersive and embodied interactions could reshape mathematical experience, representation, and classroom activity design. According to the framework of documentational genesis [8], the trainers shared their previous experience in designing XR-based educational activities. Sample activities previously implemented in upper secondary classrooms were analysed collectively to highlight how technological affordances, mathematical objectives, and pedagogical decisions shape the design of activities. Furthermore, trainers shared the theoretical perspectives that informed their design choices. In particular, embodiment theory and the theory of semiotic registers [6] were introduced to discuss how bodily interaction, movement, and multiple representations may support mathematical meaning-making within immersive environments. References to Tall's work [17] on embodied approaches to calculus were also used to frame the design of the proposed activities. Rather than presenting these frameworks in a purely theoretical way, the discussion was constantly connected to concrete classroom scenarios and to the affordances and constraints of the headsets.

Design of Classroom Activities. Teachers were first able to freely explore the apps and their features. Following this exploration, significant time was allocated to activity design sessions. These sessions served as collaborative spaces where teachers, divided into groups based on common interests and needs (such as grade level and class curriculum), could openly share ideas, pedagogical strategies and opinions. In this phase, trainers acted as facilitators, sharing their perspective, anticipating potential difficulties, and suggesting practical approaches. Teachers worked in pairs or small groups sharing a single headset. By working together, group members could guide the headset wearer using the screen mirroring feature and practice interacting with the outside world using the AR features. This setup is a practical solution to handle limited hardware availability, but was also an intentional design choice to let teachers experience the same collaborative environment their students may encounter in the classroom. This environment seems effective in allowing collaboration and interaction among group members [1]. Each group developed an activity and a lesson plan tailored to the specific needs, curriculum and composition of their own classes.

Discussion of Outcomes. The final meeting took place after a month-long break, allowing teachers to test their materials in their classes. A questionnaire was administered to evaluate the overall impact of the training program and gather feedback. Teachers shared their practical experiences and explained how the activities unfolded in the classroom. Both successful and less effective approaches were critically examined, allowing the group to weigh the pros and cons of the XR technologies, and evaluate their pedagogical impact.

5 Context and Method

Context. The program involved five upper secondary schools situated across the Piedmont region, in Northwest Italy. All participating schools feature scientific or technical curricula. In total, 35 teachers took part in the program. Three schools were selected to evaluate the impact of the PD training program. In these schools the authors of this paper acted as trainers and took part in the final discussion. Data was gathered through questionnaires administered at the end of the course in these three schools (to 18 teachers in total) and via the field notes taken by the authors throughout the program, with a particular focus on the feedback shared during the final discussion. The questionnaire included both closed and open-ended questions designed to track the implementation of activities in the classroom, and gather teachers' perceptions on the use of XR headsets and on the uniqueness of XR environments compared to a traditional approach.

Method. An inductive thematic analysis [3] was employed to qualitatively analyse the data collected from the open-ended questionnaire responses and trainers' field notes. Codes were developed organically and iteratively during the text-analysis phase to ensure the final coding scheme encompassed the full spectrum of surfaced topics. These initial codes were subsequently clustered into higher-level descriptive themes.

6 First Results and Discussion

Out of the 18 respondents, 13 implemented activities with the XR headsets in the classroom during the training program. These activities spanned all five years of upper secondary education (from 9th to 13th grade) and covered a variety of topics, including 1D kinematics, direct proportionality, derivative calculation, and potentiometric titration. Participating teachers typically utilised the headsets in the classroom between one and three times.

Through the coding process, teachers' perceived advantages of XR activities were structured into three major themes: student engagement and peer collaboration; meaning-making, visualisation and exploratory learning; and embodiment.

Student Engagement and Peer Collaboration. The most prominent theme, reported by 16 out of 18 respondents (88.9%), was a distinct shift in classroom dynamics, enhancing students' engagement and peer collaboration. With regards to engagement, a teacher observed "a higher level of emotional and cognitive engagement from the students, who were more participatory and motivated thanks to immersion in the virtual environment"[1]. According to another, the activity proved "useful for engaging even students who are not very passionate about the subject". Regarding peer collaboration, a "high level of involvement, allowing the construction of knowledge through experience, cooperative learning and inclusion" was reported. This observed collaborative student environment mirrors the design of the training program itself, where teachers worked in pairs or small groups sharing a single device to experience a cooperative dynamic.

[1] This quote and all those that follow are excerpts from questionnaire responses, translated from Italian by the authors.

Meaning-Making, Visualisation and Exploratory Learning. 14 out of 18 respondents (77.8%) highlighted how body tracking, real-time feedback, visualisation tools and the exploratory nature of the activities supported mathematical meaning-making. One teacher observed that "learning is more meaningful because they [students] 'see' the concepts being discussed", another that "the use of augmented and virtual reality also fostered a deeper understanding of the subject, especially complex or abstract topics, allowing students to visualise concepts, processes, and environments in a concrete and three-dimensional way". For instance, the use of xTimeTracer helped students "[to grasp] the difference between a space-time diagram and a trajectory, which previously was perhaps not clear to everyone". One teacher observed that the use of XR headsets allowed "to approach certain concepts through mathematical meaning rather than formalism". It is worth noting that while the theory of semiotic registers [6] was explicitly discussed during the training program, no direct mention of it emerged in the questionnaire responses. Although a multi-register approach was mentioned by some teachers during the final discussion, the visualization and exploratory capabilities afforded by XR seem to have resonated more profoundly.

Embodiment. Embodiment was explicitly recognised by 8 out of 18 respondents (44.4%) as a unique affordance of XR that cannot be easily and effectively replicated in a traditional classroom. This is not surprising, given that embodiment was one of the theoretical concepts explicitly introduced in the PD. However, responses to the questionnaires suggest that the classroom use of immersive technologies fostered a new awareness among teachers regarding the embodiment processes, enabled by AR. Specifically, one teacher highlighted that "the use of virtual reality allowed students to see in real time the translation of their bodily movement into a space-time graph". Another participant expanded on this, observing:

> The most significant and specific aspect of using the headsets, as far as I am concerned, is linked to the movement and involvement of the body. I believe it is useful for students to see that, in correspondence with certain movements, graphs are obtained from which they can deduce certain relationships.

Technical Constraints and Pedagogical Concerns. Technical and logistical issues, alongside the initial time investment required to prepare classroom activities, were the most universally cited constraints. While these concerns underscore the critical need for training programs that build technical self-efficacy, the fact that teachers actively managed these challenges during classroom activities, as emerged during the final discussion, represents a positive outcome of the program. Considering that many participants had little to no prior experience with XR headsets, the PD program seems to have empowered them to plan and implement classroom activities.

On a pedagogical level, teachers worried that without the framing of a traditional paper-and-pencil or blackboard-based lesson, students might mistake the use of XR headsets for a mere game, as such hardware is usually perceived as a recreational tool rather than a learning instrument. This preconception might result in "distraction caused by the tool itself", as reported by a teacher, and a subsequent reduction in conceptual depth. Additionally, some participants highlighted "the difficulty of linking back to the required formalization" and feared the risk of "shifting the focus onto what is seen rather than on deeper aspects", potentially leaving students underprepared for the formal approaches required in scientific high school final examinations or university entrance tests.

7 Conclusion

This study has presented the design and first implementation of a PD program aimed at supporting mathematics and physics teachers in the pedagogically meaningful integration of XR technologies into upper secondary education. Grounded in theories of embodiment, semiotic representation, and documentational genesis, the program combined technical training, collaborative activity design, classroom experimentation, and collective reflection to help teachers move beyond viewing XR as a purely technological innovation, towards using it as a resource for mathematical meaning-making.

The preliminary findings suggest that teachers perceived XR-based activities as valuable in increasing student engagement, promoting collaboration, and fostering mathematical meaning-making. Participants highlighted the value of immersive and embodied activities in helping students connect physical experience with graphical and symbolic representations, suggesting a potential contribution to conceptual understanding beyond procedural learning. At the same time, the results underscore the importance of carefully designed pedagogical mediation to ensure that XR experiences remain connected to formal mathematical reasoning and curricular goals. Technical and organizational constraints, together with concerns about balancing experiential learning and formalization, emerged as significant challenges that require continued attention.

From the perspective of documentational genesis, the collaborative design and classroom experimentation phases enabled teachers to move beyond the role of technology users and begin transforming XR applications into pedagogically meaningful resources adapted to their own instructional contexts. The PD model described here, therefore, offers a promising approach for supporting sustainable technology integration by combining technical training, theoretical grounding, collaborative design, and reflective practice.

The results reported here are preliminary and are based on a limited number of participating schools and teachers, who self-reported their experience. Future work will extend the analysis through additional data collection, including classroom observations, student learning outcomes, and longitudinal investigations of how teachers' resource systems evolve over time. Further research should also explore how XR-based embodied activities can be effectively integrated with symbolic and formal mathematical work to promote durable conceptual understanding. Overall, this study suggests that, when accompanied by carefully designed PD, immersive technologies have the potential to become a meaningful component of secondary mathematics education rather than simply an engaging technological novelty.

Acknowledgments. We are grateful to the teachers who were involved in the PD program for answering our questionnaire and sharing their precious experience. The teacher training program described in this work was funded by PAI-EDIH (Public Administration Intelligence - European Digital Innovation Hub), a European initiative dedicated to driving digital transformation and technology-driven training within public institutions and administrations. The classroom activities were developed within the CRT Foundation's Progetto Diderot and implemented by the Department of Mathematics of the University of Turin.

Disclosure of Interests. The authors have no competing interests to declare that are relevant to the content of this article.

References

1. Andriano, V., Ferrari, G.: Students' interactions in the mathematics classroom: using augmented reality to explore the derivative function. In: Clark-Wilson, A., Bokhove, C. (eds.) Proceedings of the 17th International Conference on Technology in Mathematics Teaching, pp. 32–39. London, United Kingdom (2026). https://doi.org/10.14324/000.bk.10223410
2. Ball, D.L., Thames, M.H., Phelps, G.: Content knowledge for teaching: what makes it special? J. Teach. Educ. **59**(5), 389–408 (2008)
3. Braun, V., Clarke, V.: Using thematic analysis in psychology. Qual. Res. Psychol. **3**(2), 77–101 (2006). https://doi.org/10.1191/1478088706qp063oa
4. Bulut, M., Borromeo Ferri, R.: A systematic literature review on augmented reality in mathematics education. Eur. J. Sci. Math. Educ. **11**(3), 556–572 (2023). https://doi.org/10.30935/scimath/13124
5. Cevikbas, M., Bulut, N., Kaiser, G.: Exploring the benefits and drawbacks of AR and VR technologies for learners of mathematics: recent developments. Systems **11**(5), 244 (2023)
6. Duval, R.: A cognitive analysis of problems of comprehension in a learning of mathematics. Educ. Stud. Math. **61**, 103–131 (2006). https://doi.org/10.1007/s10649-006-0400-z
7. Fried, M.N.: Meditations on future directions for AR research in mathematics education: postscript paper for touch the derivative. Digit. Exp. Math. Educ. **11**, 534–549 (2025). https://doi.org/10.1007/s40751-025-00191-y
8. Gueudet, G., Trouche, L.: Towards new documentation systems for mathematics teachers? Educ. Stud. Math. **71**(3), 199–218 (2009). https://doi.org/10.1007/s10649-008-9159-8
9. Jong, M.S.Y., Zhai, X., Chen, W.: Innovative uses of technologies in science, mathematics and STEM education in K-12 contexts. Int. J. Sci. Math. Educ. **22**(Suppl 1), 1–9 (2024). https://doi.org/10.1007/s10763-024-10530-x
10. Lakoff, G., Núñez, R.E.: Where Mathematics Comes From: How the Embodied Mind Brings Mathematics Into Being. Basic Books, New York (2000)
11. Lau, W.F.F., Jong, M.S.Y.: Typology of teachers' stages of concern for STEM education. Res. Sci. Technol. Educ. **41**(4), 1560–1578 (2023). https://doi.org/10.1080/02635143.2022.2064447

12. Lo, C.K., Ng, D.T.K., Ng, F.: Observing mathematical properties in the virtual world: an exploratory study of online independent learning of locus concepts. Int. J. Sci. Math. Educ. **22**(Suppl 1), 37–58 (2024). https://doi.org/10.1007/s10763-024-10466-2
13. Mishra, P., Koehler, M.J.: Technological pedagogical content knowledge: a framework for teacher knowledge. Teach. Coll. Rec. **108**, 1017–1054 (2006). https://doi.org/10.1111/j.1467-9620.2006.00684.x
14. Niess, M.L.: Preparing teachers to teach science and mathematics with technology: developing a technology pedagogical content knowledge. Teach. Teach. Educ. **21**, 509–523 (2005)
15. Shulman, L.S.: Those who understand: knowledge growth in teaching. Educ. Res. **15**(2), 4–14 (1986)
16. Sinclair, N.: Generations of research on new technologies in mathematics education. Teach. Math. Appl. **33**(3), 166–178 (2014). https://doi.org/10.1093/teamat/hru013
17. Tall, D.: How Humans Learn to Think Mathematically: Exploring the Three Worlds of Mathematics. Cambridge University Press, Cambridge (2013)

TuringViz: Teaching Turing Machines with Configuration Graphs and Computation Trees

Christoph Greger(✉) and Gregor Porsch

Faculty of Informatics and Data Science (FIDS), University of Regensburg, Regensburg, Germany
{christoph.greger,gregor.porsch}@stud.uni-regensburg.de

Abstract. Turing machines are a central model of computation in theoretical computer science, but their behavior can be difficult to teach: Simple local transition rules may generate complex global computations, especially for nondeterministic machines, where one input can give rise to many possible runs. We present *TuringViz.org*, a web-based educational environment for deterministic and nondeterministic multi-tape Turing machines. The tool combines concise code-like machine specifications with interactive execution and visual representations of the corresponding computation space. Its design supports lectures, tutorials, and self-study by making configurations, tape updates, and nondeterministic choices explicit and easy to inspect. Instead of showing only a single execution trace, TuringViz visualizes the reachable configurations and possible runs of a machine as configuration graphs and computation trees. This paper describes the design and functionality of TuringViz, compares it to existing visualization tools, and reports findings from an exploratory evaluation with students.

Keywords: Turing Machines · Computer Science Education · Nondeterminism · Visualization

1 Introduction

Turing machines are a central model in undergraduate courses on formal languages, computability, and theoretical computer science [8]. Standard textbooks use Turing machines to reason about decidability, computability, nondeterminism, and the limits of algorithmic problem solving [1,4,7,15]. At the same time, their behavior is not always easy to understand from the formal definition alone. Although a Turing machine is described by finitely many states, a tape alphabet, and transition rules, a concrete computation consists of a sequence of configurations that evolve over time.

This distinction between the compact machine definition and the induced computation becomes especially important for nondeterministic machines. In a nondeterministic machine, one input may give rise to several possible runs.

M. Löffler et al. (Eds.): CMSC 2026, LNCS 16951, pp. 208–221, 2027.
https://doi.org/10.1007/978-3-032-38887-2_16

Understanding such machines therefore requires keeping track of the configurations and branches generated by these transitions over time. For students, this creates a gap between the finite object written on the page, namely the transition table or state-transition graph, and the potentially large computation space induced by that object. A single execution trace shows only one sequence of nondeterministic choices, and the state-transition graph does not show which complete configurations are reachable on a fixed input. This makes nondeterministic computation a natural target for visualizations that expose branching, reachability, and complete machine snapshots.

In teaching, Turing machine computations are commonly presented using transition tables, state-transition graphs, and execution traces. These representations describe the transition relation or individual runs. This has motivated educational tools and visualizations for automata and Turing machine computations, from JFLAP [13] to more recent approaches [2,6,9,10].

We present *TuringViz*, a web-based tool for defining, executing, and visualizing deterministic and nondeterministic multi-tape Turing machines. TuringViz uses a concise domain-specific language for machine definitions and provides several synchronized views of a computation: the state-transition graph, the current tape contents during execution, the configuration graph, and the computation tree. The tool is designed for use in lectures, tutorials, and self-study, with the goal of making both the local transition behavior and the global structure of a computation visible. To the best of our knowledge, TuringViz is the first tool to provide both configuration-graph and computation-tree visualizations for Turing machines, with nodes in both views representing complete configurations.

The remainder of this paper describes the main views and interaction mechanisms of TuringViz, compares the tool to related visualization systems, and reports findings from an exploratory evaluation with students. The formal model underlying the implementation is summarized in the Appendix.

2 TuringViz as a Teaching Environment

The interface is organized into four main views: *Input*, *Run*, *Configuration Graph*, and *Computation Tree*. The input and run views support the construction and execution of concrete examples. The configuration graph and computation tree views then show how the local transition rules induce a larger computation structure.

The graph views support panning, zooming, and fullscreen display, so that they can be used both for individual exploration and as shared visual objects during lecture projection.

2.1 Specification and Stepwise Execution

The *Input* view combines a machine editor for a domain-specific language (DSL) with an automatically generated state-transition graph; see Fig. 1. The DSL

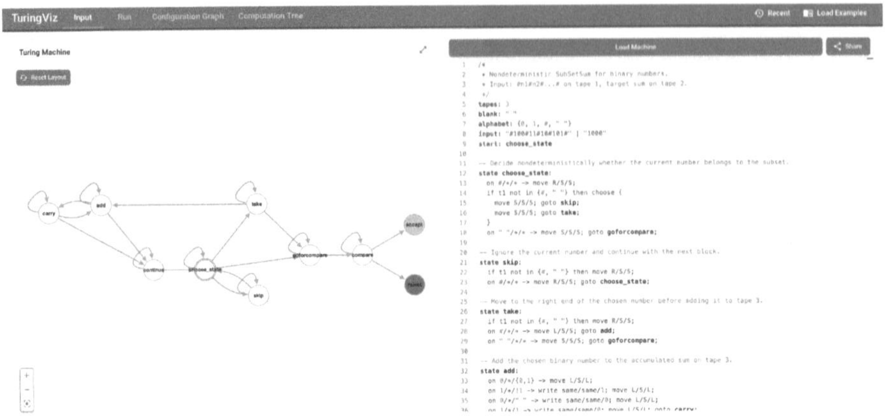

Fig. 1. Input view of TuringViz for the `NonDetSubSetSum` example. The left panel shows the generated state-transition graph, while the right panel contains the DSL editor for the machine specification.

is designed to stay close to transition-table notation while remaining compact enough for examples and live demonstrations. A machine description contains the number of tapes, the input, the blank symbol, the start state, and the transition table. Thus, students work with the same components that occur in the formal definition, but in a form that can immediately be executed.

The editor provides schema-based validation, completion, hover information, and formatting support. Thus, students receive immediate feedback while describing their intended Turing machine. When a machine is loaded, the description is parsed, checked, and used to update the simulator and all views.

TuringViz supports between one and six tapes. Practical abbreviations such as `*` for wildcard matching and `same` for preserving the currently read symbol keep examples short enough for classroom use. Lists of transition alternatives express nondeterminism explicitly.

The generated state-transition graph visualizes the transition structure of the machine. States are nodes, transitions are directed edges, and multiple transitions between the same pair of states are grouped. The start state, current state, and currently used transition are highlighted during execution. Edge tooltips show the corresponding read, write, and move actions. Accepting and rejecting states are visually marked using conventional state names such as `accept`, `done`, `reject`, or `error`.

Machines can be shared through self-contained URLs, loaded from GitHub Gists, and stored in the browser's recent-machine list after loading. These mechanisms support common teaching workflows, such as preparing examples in advance, distributing machines for exercises, or submitting homework tasks.

The *Run* view shows the current tapes together with controls for stepping, continuous execution, stopping, resetting, and adjusting the execution speed; see Fig. 2. Each tape is displayed as a horizontally scrollable row with the current head position overlaid on the corresponding cell. The view automatically centers the heads when a machine is loaded or reset, and keeps them in view during live execution. If the Turing machine is nondeterministic and there are multiple possible next configurations, the user can either choose the next configuration manually or select an automatic resolution strategy. In the latter case, TuringViz can choose a random successor configuration, follow a branch that eventually reaches an accepting configuration, or follow a branch that eventually reaches a rejecting configuration, whenever such a branch exists.

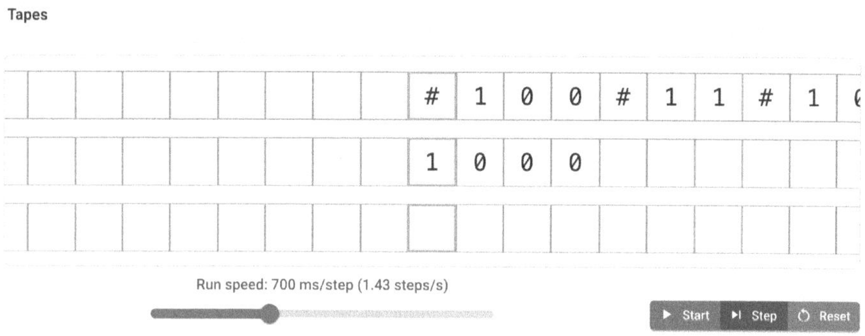

Fig. 2. Run view during the execution of a three-tape Turing machine. The view displays the current tape contents and highlights the head position on each tape. The controls below the tapes allow users to step through the computation, run it continuously, adjust the execution speed, and reset the machine.

2.2 The Configuration Graph View

The *Configuration Graph* view visualizes the reachable configurations for the currently loaded input; see Fig. 3. Each node represents a complete configuration, including the current state, tape contents, and head positions; each edge represents one computation step. Each state is assigned a distinct color and accepting and rejecting states are visually marked in the same way as in the state-transition graph.

The graph is computed after a machine has been loaded. Because computations may be infinite, graph construction is bounded by 20,000 configurations. Configurations whose successors are not fully displayed because of this limit are marked visually. Users can select any displayed configuration and set the simulator to it, making it possible to continue execution from any point in the graph.

This view visualizes configuration reachability and makes explicit that a computation is not only a sequence of control states, but a sequence of complete machine snapshots. Two configurations may share the same current state

Fig. 3. Configuration graph for the `NonDetSubSetSum` example. Each node represents a reachable configuration for the fixed input, and each directed edge represents one computation step. Nodes are colored according to their current control state; accepting and rejecting halting configurations are marked in green and red.

while differing in tape contents or head positions. Conversely, in nondeterministic machines, different paths may lead to the same configuration. The configuration graph gives instructors and students a concrete object for discussing these situations.

It also helps distinguish the state-transition graph of the machine from the computation induced by a particular input. The state-transition graph is always finite; the configuration graph is input-dependent and consists of reachable complete configurations. Showing both views in the same environment supports discussion of this distinction during a lecture or tutorial.

2.3 The Computation Tree View

The *Computation Tree* view unfolds possible runs from the initial configuration; see Fig. 4. As in the configuration graph, each node corresponds to a complete machine configuration. The root is the initial configuration, and the children of a node are the configurations reachable in one step. Unlike the configuration graph, the same logical configuration may occur multiple times in the computation tree if it is reached along different paths. As in the configuration graph view, the displayed tree is bounded, and nodes whose successors are not fully displayed because of this limit are marked visually.

This view is intended to support explanations of nondeterministic branching over time. The configuration graph represents reachability compactly, whereas the computation tree represents possible runs. A root-to-leaf path corresponds to one halting run of the machine. Accepting and rejecting leaves therefore indicate whether that run succeeds or fails.

Both the configuration graph and the computation tree can be shown in compact node mode or in a card mode that embeds tape snapshots directly into

Fig. 4. Computation tree for the `NonDetSubSetSum` example. The root represents the initial configuration, and each root-to-leaf path corresponds to one possible finite run. Unlike the configuration graph, the computation tree keeps different nondeterministic branches separate, even if they reach the same configuration.

the graph; Fig. 5 shows the latter mode for an example. For larger computation trees, deterministic linear segments can be compressed so that the visualization emphasizes actual branch points. These options allow the same example to be adapted to different teaching goals: compact node mode supports overview and discussion of structure, while card mode supports closer inspection of individual configurations.

2.4 Implementation Details

The configuration graph and computation tree are generated using breadth-first search. To detect previously visited configurations, each configuration is serialized into a canonical representation and stored in a JavaScript `Map`, ensuring that each configuration appears only once in the configuration graph. Graph layouts are computed using ELK.js [5] for both visualization modes. The resulting layouts are rendered using React Flow [17] in card mode and Cytoscape.js [3] in node mode, where WebGL-based rendering enables responsive visualization of large graphs. To avoid partially expanded configurations when the visualization limit is reached, successor configurations are generated atomically: either all successor configurations of a configuration are added or none are. Consequently, the default limit of 20,000 configurations may be exceeded slightly in order to complete the expansion of the final processed configuration. This limit was chosen because larger graphs noticeably reduce the responsiveness of interactive visualizations on typical hardware.

2.5 Using the Views Together

The four views are intended to support a common lecture pattern: moving from a formal transition rule to its effect on a concrete configuration and then to its

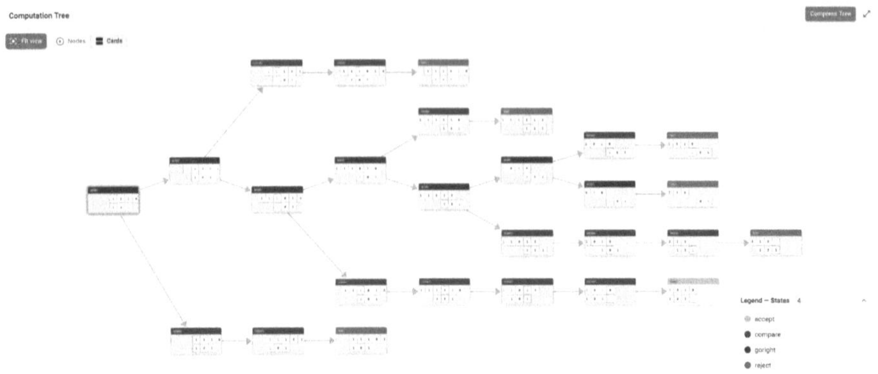

Fig. 5. Computation tree view for the `vvWord` example in card mode. Nodes show complete configurations directly as embedded tape snapshots, while branches represent nondeterministic choices during the computation.

role in the global computation. An instructor can present or modify the transition table, ask students to predict the next configuration in the run view, and finally use the configuration graph or computation tree to discuss reachability, branching, or accepting paths.

This workflow makes it possible to discuss the same transition at several levels of abstraction: as part of the finite control, as an operation on the current tapes, as an edge between reachable configurations, and as one step in a possible run. In this way, TuringViz connects the compact formal description of a machine with the computation structure it induces.

2.6 Built-in Examples and Explanatory Support

TuringViz includes built-in examples for common lecture scenarios: `BinaryAdd`, `NonDetSubstring`, `vvWord`, `NonDetSubSetSum`, `CheckEven`, `GCD`, `AllStrings`, `Fib`, `DAG`, `Circle`, and `NonDetSAT`. These examples illustrate a broad spectrum of Turing machine computations, ranging from simple deterministic procedures to nondeterministic search problems such as subset sum and SAT.

The website also provides contextual information for the different views. The footer updates with the active tab and provides concise explanations of the relevant concepts, including Turing machines, configurations, configuration graphs, computation trees, and the DSL specification.

3 Related Work

The closest web-based tools to TuringViz are turingmachine.io [16] and the Turing machine simulator TUR [12]. Both provide a browser-based interface for writing Turing machine descriptions, visualizing the state-transition graph, and executing a machine on a concrete input. TUR also supports multi-tape Turing machines. In both tools, however, the visualization is limited to the machine description and concrete execution: they do not support nondeterministic Turing machines and do not provide configuration graphs or computation trees. TuringViz follows the same web-based approach, which lowers the entry barrier for students because no installation is required and examples can be opened directly from shared links. This also makes the tool usable on devices such as tablets.

The Gidayu tool is closely related to TuringViz in its treatment of computations [2]. It visualizes computations by means of so-called computation diagrams whose nodes are configurations and whose edges represent computation steps. In this respect, Gidayu provides a similar kind of visualization to the configuration graph in TuringViz. The difference is the supported machine model: Gidayu targets automata such as nondeterministic finite automata and nondeterministic pushdown automata, whereas TuringViz applies this idea to deterministic and nondeterministic multi-tape Turing machines.

Morazán et al. describe computation graphs in their FSM tool [10]. Their notion of a computation graph differs from the configuration graphs used in TuringViz: in FSM, a computation graph is a subgraph of the state-transition graph consisting of the states and transitions used during a computation. Thus, its nodes are states rather than complete configurations. In TuringViz, nodes of the configuration graph are configurations, including the current state, tape contents, and head positions. The FSM work also addresses the problem that computations may be infinite. As in TuringViz, graph generation is bounded by a threshold, and the user is informed when the displayed graph may be incomplete. Recent work on the FSM tool adds support for multi-tape Turing machines [9].

JFLAP [13,14] is a broader educational environment for formal languages and automata theory and also supports nondeterministic multi-tape Turing machines. Its execution visualization for nondeterministic machines differs from the one used in TuringViz. During stepwise execution, JFLAP maintains a list of current configurations; after one step, this list is replaced by the list of all successor configurations. This makes the set of currently active branches visible during simulation. However, JFLAP does not present the computations as a configuration graph or as a computation tree.

4 Evaluation

As an initial step toward evaluating its educational usefulness, an earlier prototype of TuringViz was used in the course *Complexity Theory* at the University of Regensburg, where it served both as a visual support tool in lectures and as the basis for student exercises. This prototype already supported interactive execution and visualization of Turing machine computations, but used a YAML-based machine specification format rather than the DSL described in this paper. In the exercises, students specified Turing machines using this earlier format, and their solutions were subsequently discussed in tutorials. This initial classroom use led to positive informal feedback, but was not accompanied by structured data collection.

To complement these observations with systematic feedback, we conducted a guided exploratory evaluation with six undergraduate computer science students, all in their sixth semester of study. Three participants reported no prior knowledge of Turing machines, one reported basic familiarity, and two reported moderate understanding from previous coursework. The session combined brief theoretical introductions, individual interaction with example machines, and two implementation tasks. The latter were included because constructing machines requires participants to engage with the website more actively than merely inspecting predefined examples. Whether participants were able to complete these tasks therefore provided a qualitative indication of whether they could use TuringViz to construct machines and reason about their behavior.

First, participants received a brief explanation of deterministic Turing machines, including their basic operational semantics. They then worked with the built-in `BinaryAdd` example to familiarize themselves with the interface, DSL specification, state-transition graph, and stepwise execution. Afterwards, they implemented a deterministic two-tape machine deciding the language

$$L_1 = \{\, v\#v \mid v \in \{0,1\}^* \,\}.$$

The second phase followed the same structure, but shifted the focus to nondeterminism. After an introduction to nondeterministic Turing machines, participants interacted with the built-in `vvWord` example before implementing a nondeterministic one-tape machine deciding the language

$$L_2 = \{\, w \in \{0,1\}^* \mid \exists i \in \{1,\ldots,|w|-3\} : w_i = 1 \land w_{i+3} = 1 \,\}.$$

This language was chosen because it can be decided using a simple nondeterministic strategy: the machine nondeterministically guesses a position in the input and verifies locally that the symbols at distance three are both equal to 1. The example therefore illustrates the central idea of nondeterministic guessing followed by deterministic verification while keeping the machine description compact. Finally, the students explored the larger `NonDetSubSetSum` example. All participants solved both implementation tasks successfully, with varying completion times.

After the session, participants answered a questionnaire with eight five-point Likert-scale items, where 1 corresponded to "strongly disagree" and 5 to "strongly agree". The questionnaire addressed perceived clarity, usability, and support for learning. It also included two free-text questions about positive aspects and possible improvements.

The response distribution in Fig. 6 indicates a positive initial reception. Across all eight items, 45 of 48 responses were either "agree" or "strongly agree". The strongest agreement concerned the understandability of the execution view, the usefulness of stepwise nondeterministic execution, and the willingness to recommend TuringViz. The lowest-rated item was the interpretation of colors and symbols.

The free-text responses were consistent with these results. Participants particularly valued the stepwise execution, tape visualization, and the possibility

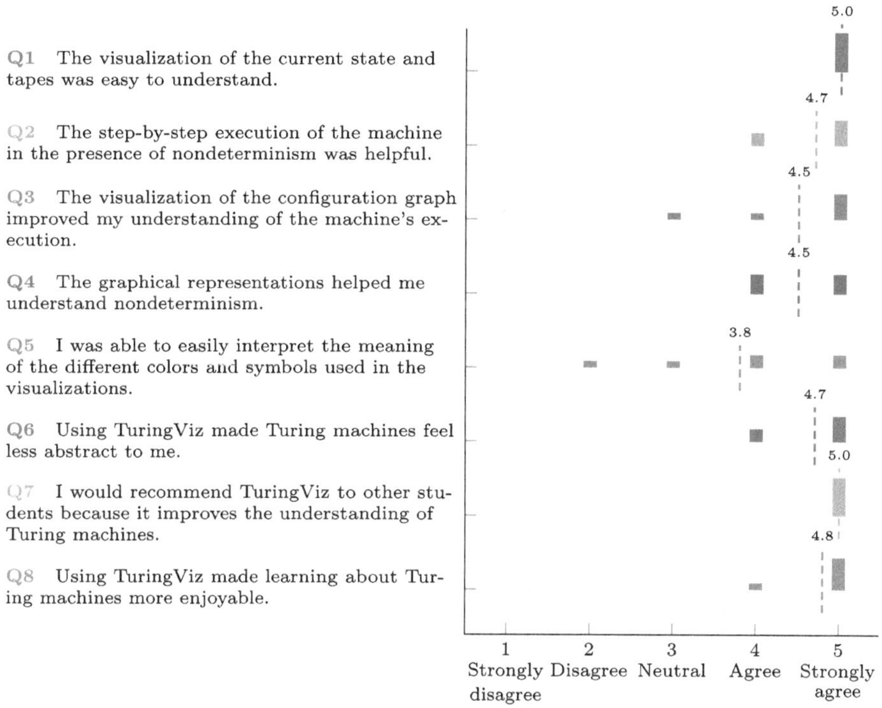

Fig. 6. Response distributions for the Likert-scale questions in the exploratory evaluation of TuringViz. Each question was answered by all six participants. Dashed vertical lines indicate the corresponding mean response.

to construct and test machines themselves. For instance, students praised the "execution animation" and highlighted that the tool provides a "graphical understanding for an abstract concept." Suggested improvements mainly concerned interface guidance and interaction flexibility. Participants asked for clearer support when learning the DSL syntax, clearer explanations of what the colors and symbols in the visualizations denote, and direct editing of tape inputs within the active run view.

Given the scope of the study, these results should be interpreted with appropriate caution. The evaluation was exploratory and involved only six participants. The successful completion of both implementation tasks provides qualitative evidence that participants were able to use TuringViz for the intended activities. However, the questionnaire primarily measured perceived usefulness and usability rather than actual learning gains. We therefore do not claim that TuringViz improves learning outcomes in comparison with traditional teaching materials. Instead, the results suggest that students perceived the visualizations as understandable and useful.

5 Conclusion and Future Work

TuringViz bridges the gap between compact Turing machine specifications and the complex computation spaces they induce. By integrating a concise DSL with stepwise execution, configuration graphs, and computation trees, the tool enables students and instructors to inspect local rule applications and global execution paths within a single synchronized environment. Our preliminary evaluation indicates that students found this interactive and visually clear representation helpful, while also highlighting areas for improvement – particularly regarding the readability of symbols and color choices – which we plan to address in future iterations.

Beyond these evaluation-driven improvements, future work will focus on scalability and analytical support. Because even simple machines can generate very large computation spaces, we plan to improve graph generation, rendering performance, and layout behavior for extensive configuration graphs and computation trees. In addition, TuringViz could make recurring structures easier to recognize, for example by detecting and highlighting repeated subtrees or sequences of configurations. A sidebar could list the most frequent patterns and allow users to inspect the corresponding parts of the graphs directly.

We also plan to explore alternative renderings of configuration graphs beyond the current DAG-oriented layout, in order to make cycles, repeated configurations, and structural regularities more visible. A further direction is to reduce the effort required to define larger machines. Inspired by work on composed Turing machines [11], subroutine-like abstractions could allow users to reuse common machine fragments instead of specifying every transition manually. Finally, larger-scale quantitative studies are needed to assess the pedagogical impact of TuringViz more robustly and to compare it with more traditional forms of teaching Turing machine computations.

Acknowledgments. We thank Radu Curticapean for supervising the initial student project from which TuringViz originated and for his feedback and support during its further development.

Disclosure of Interests. The authors have no competing interests to declare that are relevant to the content of this article.

Appendix: Formal Model

We use a standard formalization of Turing machines [1,4,7,15] with minor adaptations and describe the execution of Turing machines in terms of configurations, following the notion used in [1].

Turing Machines. A deterministic k-tape Turing machine is a tuple

$$M = (Q, \Sigma, \Gamma, b, k, \delta, q_0),$$

where

- Q is a finite set of *states* and $q_0 \in Q$ is the *start state*,
- $\Sigma \subseteq \Gamma \setminus \{b\}$ is the *input alphabet*, Γ is the *tape alphabet*, and $b \in \Gamma$ is the *blank symbol*,
- $k \in \mathbb{N}$ is the number of tapes, and
- δ is the partial *transition function*

$$\delta : Q \times \Gamma^k \rightharpoonup Q \times \Gamma^k \times \{L, S, R\}^k.$$

A nondeterministic k-tape Turing machine is defined analogously, except that the partial transition function maps to sets:

$$\delta : Q \times \Gamma^k \rightharpoonup \mathcal{P}(Q \times \Gamma^k \times \{L, S, R\}^k).$$

Transition Function. The transition function δ receives the current state and the k tape symbols under the heads. Each possible transition then specifies (i) the next state, (ii) the symbols to be written to the currently read tape cells, and (iii) the movement direction of each head. We use the movement convention

$$L = -1, \qquad S = 0, \qquad R = +1,$$

to denote the head movements.

Configurations. A configuration of a k-tape Turing machine is a complete snapshot of the machine at one moment in time. We write configurations as triples

$$C = (q, h, T) \in Q \times \mathbb{Z}^k \times B^k.$$

Here

- $q \in Q$ is the current state,
- $h = (h_1, \ldots, h_k) \in \mathbb{Z}^k$ is the vector of head positions, and
- $T = (T_1, \ldots, T_k) \in B^k$ is the vector of tape contents.

The set B consists of all finite partial tape contents

$$B = \{T : \mathbb{Z} \rightharpoonup \Gamma \mid T(j) \text{ is defined for only finitely many } j \in \mathbb{Z}\}.$$

Initial Configuration. For an execution in TuringViz, the input is a k-tuple $x = (x_1, \ldots, x_k)$ of words over the input alphabet, with $x_i \in \Sigma^*$ for each $1 \le i \le k$. Initially, x_i is written on tape i, beginning at position 0 and continuing to the right. Formally, for each tape i, we define the initial tape content $T_i^x : \mathbb{Z} \rightharpoonup \Gamma$ by

$$T_i^x(j) = x_i[j] \text{ for } j \in \{0, \ldots, |x_i| - 1\} \qquad \text{and} \qquad T_i^x(0) = b \text{ if } |x_i| = 0$$

With $h_0 = (0, \ldots, 0)$ and $T^x = (T_1^x, \ldots, T_k^x)$, the initial configuration for input x is

$$C_0(x) := (q_0, h_0, T^x).$$

This differs from the usual textbook convention, where the input is placed only on the first tape and all other tapes are initially blank. The difference is only a visualization convenience: the tuple $(x_1, \ldots, x_k)$ could be encoded on the first tape using delimiters and then copied to the remaining tapes by the machine. Users who want the traditional convention can leave the additional tapes blank.

Single-Step Relation. Let $C = (q, h, T)$ be a configuration. A configuration $C' = (q', h', T')$ is a successor of C, written $C \Rightarrow_M C'$, if the transition function contains a matching possible transition for the current state and the current read symbols. To obtain C', the specified symbols are first written to the current tape cells, the heads are then moved according to L, S, or R, and the machine finally enters the new state q'. If a head moves to a tape cell that has not been defined in T before, this cell is added to T' with value b. Thus T' contains the tape cells initialized by the input and any additional cells visited during the computation, while all other cells are interpreted as blank.

If no transition is applicable in configuration C, then C has no successor and the machine halts. If several transitions are applicable, then C has several possible successors. This is precisely the source of nondeterministic branching.

Executions. An execution, or run, of M on input x is a sequence of configurations

$$C_0(x) \Rightarrow_M C_1 \Rightarrow_M C_2 \Rightarrow_M \cdots$$

starting from the initial configuration and following the one-step relation in every step. A finite run ends in a configuration without successors. For nondeterministic machines, the same input may induce many different runs.

Configuration Graph. For a fixed machine M and input x, the configuration graph $G_M(x)$ is the directed graph whose nodes are exactly the configurations reachable from the initial configuration and whose edges correspond to single computation steps. Formally,

$$G_M(x) := (V, E), \qquad V := \{C \mid C_0(x) \Rightarrow_M^* C\}, \quad E := \{(C, C') \mid C \Rightarrow_M C'\}.$$

Here $\Rightarrow_M^*$ denotes reachability in zero or more steps. The distinguished start node of the graph is $C_0(x)$.

Intuitively, each node of the configuration graph is a complete machine snapshot, and each directed edge represents one computation step. In a deterministic machine, every node has at most one outgoing edge. In a nondeterministic machine, a node may have several outgoing edges. Nodes without outgoing edges correspond to halting configurations.

Computation Tree. The computation tree is the unfolding of the configuration graph from the root configuration $C_0(x)$. The root is the initial configuration, and the children of a node are exactly the configurations reachable in one step from that node. If the same logical configuration is reached along different paths, it appears several times in the computation tree, because the tree represents possible runs rather than only reachability. Leaves of the computation tree are therefore halting configurations, and a path from the root to a leaf corresponds to one finite run of the machine.

References

1. Arora, S., Barak, B.: Computational Complexity: A Modern Approach. Cambridge University Press, Cambridge (2009). https://doi.org/10.1017/CBO9780511804090
2. Cogumbreiro, T., Blike, G.: Gidayu: visualizing automaton and their computations. In: Proceedings of the 27th ACM Conference on Innovation and Technology in Computer Science Education, pp. 110–116. ACM, New York (2022). https://doi.org/10.1145/3502718.3524742
3. Franz, M., Lopes, C.T., Huck, G., Dong, Y., Sumer, O., Bader, G.D.: Cytoscape.js: a graph theory library for visualisation and analysis. Bioinformatics **32**(2), 309–311 (2016). https://doi.org/10.1093/bioinformatics/btv557
4. Hopcroft, J.E., Motwani, R., Ullman, J.D.: Introduction to Automata Theory, Languages, and Computation, 3rd edn. Addison-Wesley, Boston (2006)
5. Kasperowski, M., Domrös, S., von Hanxleden, R.: The Eclipse Layout Kernel. In: Proceedings of the 32nd International Symposium on Graph Drawing and Network Visualization, GD 2024, pp. 56:1–56:2. Schloss Dagstuhl – Leibniz-Zentrum für Informatik, Dagstuhl (2024).https://doi.org/10.4230/LIPIcs.GD.2024.56
6. Kempinski, O., Morazán, M.T.: Visualizing why nondeterministic finite-state automata reject. In: Presented at the Scheme and Functional Programming Workshop 2023 (2023). https://doi.org/10.48550/arXiv.2310.08025
7. Kozen, D.: Automata and computability, 1st edn. In: Undergraduate Texts in Computer Science. Springer, Heidelberg (1997). https://doi.org/10.1007/978-1-4612-1844-9
8. Kumar, A.N., et al.: Computer Science Curricula 2023. ACM, New York (2024). https://doi.org/10.1145/3664191
9. Morazán, M.T., Kempinski, O., Garced, A.M.: Design support for multitape Turing machines. Electron. Proc. Theor. Comput. Sci. **424**, 1–24 (2025). https://doi.org/10.4204/EPTCS.424.1
10. Morazán, M.T., Kempinski, O., Garced, A.M.: The implementation of FSM computation graphs. In: Proceedings of the 36th Symposium on Implementation and Application of Functional Languages, pp. 13–24. ACM, New York (2025). https://doi.org/10.1145/3723325.3723337
11. Morazán, M.T., Minić, T., Kempinski, O.: Visualizing composed Turing machines. In: Proceedings of the 2024 ACM SIGPLAN International Symposium on SPLASH-E, SPLASH-E 2024, pp. 34–44. ACM, New York (2024). https://doi.org/10.1145/3689493.3689979
12. Rezigned: Turing machine language simulator. https://rezigned.com/tur/. Accessed 20 July 2026
13. Rodger, S.H., Bressler, B., Finley, T., Reading, S.: Turning automata theory into a hands-on course. In: Proceedings of the 37th SIGCSE Technical Symposium on Computer Science Education, SIGCSE 2006, pp. 379–383. ACM, New York (2006). https://doi.org/10.1145/1121341.1121459
14. Rodger, S.H., Wiebe, E., Lee, K.M., Morgan, C., Omar, K., Su, J.: Increasing engagement in automata theory with JFLAP. In: Proceedings of the 40th ACM Technical Symposium on Computer Science Education, SIGCSE 2009, pp. 403–407. ACM, New York (2009). https://doi.org/10.1145/1508865.1509011
15. Sipser, M.: Introduction to the Theory of Computation, 3rd edn. Cengage Learning (2012)
16. TuringMachine.io: Turing machine visualization. https://turingmachine.io/. Accessed 20 July 2026
17. XYFlow: React Flow. https://reactflow.dev/. Accessed 20 July 2026

Strategies in Geometry Problem Solving Supported by Dynamic Geometric Software

Andreas Imhof and Matthias Müller(✉)

University of Teacher Education of the Grisons, Scalärastr. 17, 7000 Chur, Graubünden, Switzerland
{andreas.imhof,matthias.mueller}@phgr.ch

Abstract. Little is known of how Dynamic Geometric Software (DGS) are used by learners to engage geometric problem solving. Especially what kind of problem solving strategies are triggered using DGS and how learners' cognitions could be measured unbiasedly. In this context it is important to consider different cognitive processes involved in the measurement, especially if they are more recall or recognition driven. Therefore, a thinking-out-loud study is conducted, where learners communicated in a one-to-one interview setting about their procedures during solving a geometrical problem, that usually lie outside their zone of proximal development. During the process, digital aids, in concrete tailored GeoGebra tools, are provided stepwise to learners. With every digital aid provided, there is more potential for memory recall of known, but at this time not considered, problem solving strategies. In total 23 interviews with skilled Swiss primary school learners were conducted, transcribed and subsequently coded by a mostly deductively derived coding manual containing main- and subcodes containing stage, problem solving strategies and Problem Solving Cycle. For analysis a content analysis is conducted with focus on stage (tailored digital aid) and problem solving strategy as comparing areas, slopes or counting edges. The article discusses an individual example combining the codes on strategies in dependence of DGS provided aids. First results show that digital aids have the potential to trigger adequate problem solving strategies.

Keywords: Problem Solving Strategies · Recall and Recognition · Tailored Digital Aids · Dynamic Geometry Software

1 Introduction

In mathematics educational research, digital mathematical tools are intensively studied in the teaching and learning of mathematics. However, there is a lack of research on the qualitative use of digital mathematical tools in this field (Ostermann et al., 2021). Digital mathematical tools such as Dynamic Geometry Software (DGS) are considered to have noticeable potential with respect to conceptual understanding: they may help learners to gain conceptual as well as procedural knowledge when working on mathematical problems (Barzel, Büchter, & Leuders, 2007, p. 70). The potential for transfer among problem solving strategies is of particular interest (Heinrich, Bruder, & Bauer, 2015). So

M. Löffler et al. (Eds.): CMSC 2026, LNCS 16951, pp. 222–237, 2027.
https://doi.org/10.1007/978-3-032-38887-2_17

far, research questions specifically addressing problem solving on geometry topics while using DGS are rare. Current research on geometric problem solving focuses mainly on relevant heuristics and phases of the problem solving process (Rott et al., 2021).

The goal of this qualitative study is therefore to understand problem solving strategies of learners evoked by provided digital aids, based on an exemplary geometric problem using DGS. The geometric problem involves two geometric arrangements (a) and (b), perceptually resemble triangles but are in fact polygons (see Fig. 1): Each arrangement is composed of the same plane figures; however, due to different slopes of the figures there is seemingly a "missing" square in one arrangement. This may cause a cognitive conflict for learners. Without any aids this problem lies for young learners on primary level out of their zone of proximal development (Vygotsky, 1978). DGS then offers learners the potential to compare the arrangements and test hypotheses of why they are not the same. Cognitive processes are not directly observable, and (basic) mathematical concepts could only be inferred by analyzing learning artifacts, such as written tasks, observations during problem solving and reflecting interviews with learners. Each of these methods has its own epistemological advantages and disadvantages as well as method-specific biases (Hasselhorn, & Schneider, 2007). In addition, different testing methods can activate different cognitive processes (e.g., recall and recognition). This is the reason for the stepwise introduction of tailored digital aids to learners during problem solving in the study. The stepwise study design aims to move from mostly memory-based recall to more recognition-based thinking. Learners were asked to think aloud within standardized qualitative interviews. For each aspect (1) cognitive processes such as recall and recognition, (2) mathematical Problem Solving Cycle and (3) verbalization of cognitive processes with thinking-out-loud a theoretical approach was chosen to alight the study goal.

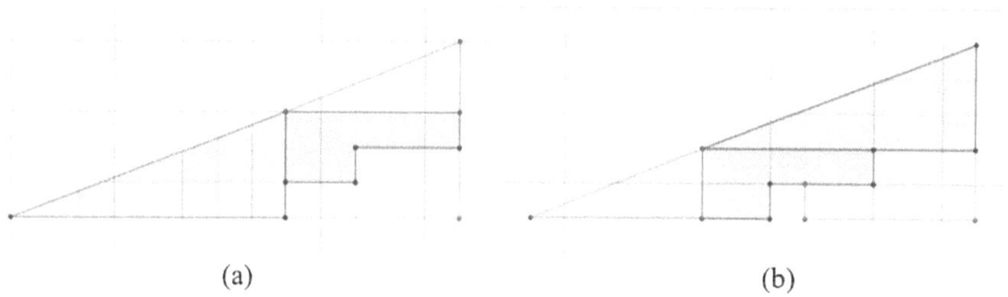

Fig. 1. Two arrangements (a) and (b) of four geometric figures (Müller, 2024). Created with GeoGebra (http://www.geogebra.org), licensed under CC BY-NC-SA 0 (CC BY-NC-SA 3.0 Legal Code | Attribution-NonCommercial-ShareAlike 3.0 Unported | Creative Commons).

2 Theoretical Approach

As mentioned above, the present study aims to understand problem solving strategies of learners within a geometric problem evoked by providing digital aids by a DGS. Since students' thinking processes such as problem solving strategies are of a latent nature, appropriate methods must be found to diagnose them as undistorted as possible. The result of the measurement can depend on which survey methods are used to investigate learners' cognitive processes. In diagnostics, the results of measurements can depend on whether the task stimulates recognition processes or rather recalls information from memory. Each approach potentially producing different biases (underestimation or overestimation of learners' knowledge) in the results (Müller, & Imhof, 2025; Imhof, & Kübler, 2025). Following this consideration leads to the theoretical approach of recall and recognition paradigm.

2.1 Recall and Recognition Paradigm

Beginning with the fact that measurement influences itself, the (measuring) instrument is also simultaneously the object being examined: knowledge or the memory of the learners. Research must therefore consider that every measurement of learners' knowledge, especially for children, involves a distorting "*bias*". In other words, that valid measurements cannot be assumed. It seems even more important that research becomes aware of the directions in which different methods can distort results under certain circumstances. Of particular relevance in this regard appears the distinction of whether diagnostic tasks elicit processes of information retrieval from memory (recall paradigm) or rather processes of recognition (recognition paradigm). By "*recall*" it is meant that tasks are designed in such a way that people must remember information learned in the past (Raaijmakers, & Shiffrin, 1992, pp. 207). This information must be freely retrieved from episodic memory (Baddeley, 2001). An example of this is the prompt to remember and write down words from a previous learning phase. In psychology, this is distinguished from processes of "*recognition*". In this case, one or more additional aides are provided in the task, which make it possible to compare the aid with the information in episodic memory (Raaijmakers, & Shiffrin, 1992, p. 208). An example of this is that subjects have to decide whether a shown word occurred in an earlier learning phase or not. Evidence that recall and recognition are different cognitive processes anchored in memory is well documented in empirical research (Haist, Squire, & Shimamura, 1992) and also seems to be confirmed in clinical practice with different lesions in memory, as patients, for example, are unable to perform tasks in the sense of the recall paradigm but show normal performance in recognition (e.g., Elward et al., 2024).

Since in both paradigms the same information in episodic memory is under investigation, both paradigms should lead to similar results. In empirical studies, however, it is repeatedly shown that test subjects perform better on tasks in the recognition paradigm compared to the recall paradigm (e.g., Rhodes, Greene, & Naveh-Benjamin, 2019; Tversky, 1973). With respect to the present study, an adaptive design was chosen, in which stepwise tailored digital aids were introduced to learners during problem solving. In doing so, the two paradigms of the theoretical approach are determined according to recall and recognition. Tailored digital aids regarding the possibility of stimulating problem solving strategies are located within the two paradigms (see Fig. 2). Problem solving strategies are located within the Problem Solving Cycle, which leads to second theoretical approach, which was considered according to the study goal.

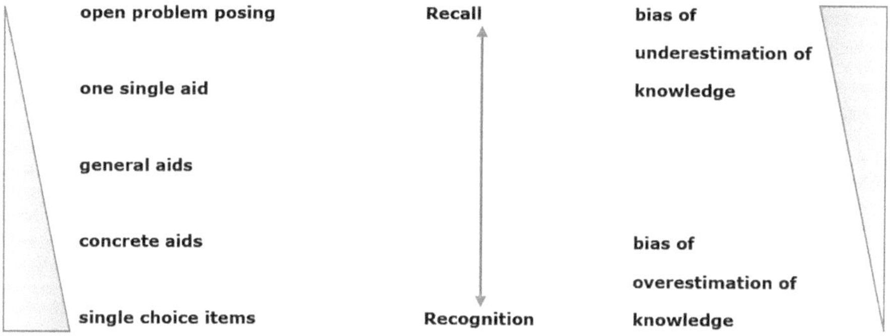

Fig. 2. Tailored aids among the theoretical approaches of recall and recognition paradigm with possible biases.

2.2 Mathematical Problem Solving Cycle

Regarding problem solving, the possibility of transfer potentials under problem solving strategies is of particular interest (Heinrich, Bruder, & Bauer, 2015). Research with the focus specifically on geometric problems in connection with the use of DGS requires further investigation. The current research focus in this field is mainly on corresponding heuristics and models of the Problem Solving Cycle (Rott et al., 2021).

With respect to Pólya (1954) posing a problem can engage problem solving, which stimulates an iterative process: The Problem Solving Cycle consist of four phases: (a) understanding the problem, (b) making a plan, (c) carrying out the plan and (d) looking back.

(a) "*Understanding the problem*" means the posed problem can be reproduced in learner's own words. Moreover, the problem is understood in such a way that a cognitive conflict has been triggered within the learners, which leads to an insight into the necessity of productive engagement with the problem (Fernandez et al., 1994).
(b) "*Making a plan*" describes that a procedure for tackling the problem is drawn up. Therefore, a problem solving strategy is initiated. This may involve selecting one problem solving strategies among different (Fernandez et al., 1994).
(c) "*Carrying out the plan*" describes a phase, where learners work on the plan of phase (b), meaning a problem solving strategy is clear. If no immediate solution occurs, the plan can be adjusted, which leads to phase (d) and may include a change of an adequate problem solving strategy (Fernandez et al., 1994).
(d) "*Looking back*" is an evaluation of the chosen procedure. The results of the problem solving strategy are compared with the problem conditions. If a solution is appropriate, it can be justified. If a difference from the problem posing occurs, the process is constructively reflected: The plan can be adjusted and a different problem solving strategy can be chosen (Fernandez et al., 1994).

Problem based learning is often described in connection with the use of DGS but are rarely observed in practice. Crucial is the coordination of use of DGS regarding individual learning processes (Ostermann et al., 2021) such as the Problem Solving Cycle.

A judgment about the quality of the use of DGS in mathematical problem solving is generally difficult (Müller, 2024). The goal of the present study is therefore to understand problem solving strategies of learners evoked by provided digital aids, based on an exemplary geometric problem using DGS. To pursue the formulated study goal, the method of think-out-loud according to Konrad (2010) was chosen to document the cognitive processes of learners. Since the study focuses on problem solving with DGS, it was considered that the change in learners' cognitive performance due to verbalization of thoughts may have an influence on the study's results (see Sect. 2.1). That is the reason while the theoretical approach (3) of the think-out-loud needs to be reflected in more detail.

2.3 Thinking-Out-Loud

According to Konrad (2010, p. 476), it can be stated that thinking-out-loud is initially characterized by the fact that learners are required to formulate an action-accompanying, oral description of their cognitive planning and procedures. In this procedure, learners are prompted in a one-to-one interview with an interviewing person to verbalize their thoughts while completing a problem. The approach belongs to the empirically inductive verbalization procedures, as qualitative data are used to infer cognitive processes in learners in general. Thinking-out-loud is a summary of different types (Ericsson, & Simon, 1998); in this context, a further distinction can be made: Speaking out loud ("talk aloud") refers to the articulation of information that is already in verbal form, such as the verbalization of applied strategies. Thinking-out-loud ("think aloud") refers, in a narrower sense, to the articulation of information that still needs to be verbalized. A translation effort is required for this. Both types can be supplemented by tailored aids ("concurrent probing"), which refers to the targeted prompt for verbalization (Ericsson, & Simon, 1998). Similarly, different types of thinking-out-loud can be described regarding the temporal engagement with learner's own thinking: "*Introspection*" refers to the ability to talk about current thoughts, "*immediate retrospection*" describes retrospective self-observation, and "*delayed retrospection*" involves an additional time delay (Konrad, 2010, p. 476). The type of introspection is the closest connection between thinking and verbal reports when learners put their thoughts into words immediately during problem solving. It can be aligned with think-out-loud protocols according to Ericsson, & Simon (1998).

With thinking-out-loud, different goals can be pursued. One goal can be the description of cognitive processes (Börnert, & Wilbert, 2015). Another goal can be the support of cognitive processes in terms of self-construction. Accordingly, thinking-out-loud is used to comprehend or support different cognitive processes. Thus, the processing of content can be verbalized by addressing the processing of information, which includes explicitly problem solving and drawing connections such as by aids evoked strategies (Fahy, 2014).

3 Research Questions

Against the background of the theoretical approaches the following research questions (RQ) were formulated:

1. Which problem-specific strategies can be identified when working on the geometric problem?
2. Which problem-specific strategies are evoked by which digital aids of DGS?

These questions guided the study in an exploratory sense and determined the chosen methods.

4 Methods

4.1 Geometric Problem with DGS

The given geometric problem involves two arrangements (a) and (b) of four figures (see Fig. 1). Both arrangements resemble a triangular shape; however, the supposed triangle in arrangement (b) includes an additional square. This initially leads to an apparent contradiction regarding the area. The task therefore aims at identifying the relation of the square from arrangement (b) in combination to arrangement (a). Arrangement (a) corresponds, as a whole figure, to a concave quadrilateral, and arrangement (b) corresponds, as a whole figure, to a concave irregular octagon (Müller, 2024).

4.2 Sample

Overall, 23 learners worked on the geometric problem and were surveyed within standardized one-to-one interviews. The average age was 10.95 years (SD = 1.496 years) and ranged from 7 to 14 years. There were 5 female and 18 male learners in the sample. All learners were participants of one of the mathematics and computer science enrichment programs conducted by the University of Teacher Education of the Grisons and therefore supposed to be above-average interested in mathematical and computer science topics.

4.3 Setting

Learners were interviewed in a one-to-one setting, either one of the two authors or two student assistants conducted the interview according to a standardized field manual. The procedure was planned in multiple stages: (1) learners were welcomed and introduced to the study goals, (2) they were made familiar with the DGS GeoGebra by solving three simple tasks (drawing a line, a triangle and measuring an area) and (3) with the thinking-out-loud method by solving a simple multiplication ("8 times 12") while talking about the individual approach to do the calculation. After this initial proceeding (4) the actual task started by requesting the pupils to read the problem posing (see Sect. 4.1). At this stage, no further aid was provided to the learners. Then, when the learners made no visible progress, the aids were activated stepwise in the following order: (5.1) Zoom-tool and line-drawing-tool were highlighted, (5.2) a reference line in both arrangements occurred, (5.3) the area number for the "missing" square and the areas over and under the reference line occurred. If the learners then could not develop a problem solving strategy, interviews were stopped and (6) learners were asked to describe their experiences with the problem, GeoGebra and the thinking-out-loud procedure. Finally, (7) a description of the solution was explained to the learners (off-record).

During the problem solving phases the field manual made the following suggestions for the conduct of the interviews: asking the pupils to explain their actions; asking open questions; no correction of false strategies and solutions; accepting nescience; usage of encouraging statements ("you are doing very well"); no time pressure, giving time to think and explain. These procedures should first ensure objectivity in the sense of not affecting learners' knowledge by interviewers influence and second guarantee a comfortable testing environment for learners to increase their speaking output. As Ericson, & Simon (1998) suggested, different strategies of concurrent probing to enhance pupils' speaking are applied:

- To retrace: Ask the pupils to explain what they do when they are silent at work or thinking.
- Explanation: Encourage pupils to make assumptions, deliberate adequate problem solving strategies.
- Encouragement: Showing respect for thoughts and ideas (e.g. "that's interesting", "could you tell me more about it").

4.4 Categories and Coding Manual

The coding manual was first developed in a structural and deductive manner: To distinguish the different stages of digital aids during problem solving six stages were defined: (S1-S3) introduction (explanations and examples), (S4) problem posing without aids, (S5.1) highlighting DGS-tools "zoom" and "straight line", (S5.2) appearance of reference lines (in both arrangements), (S5.3) appearance of fulfilling areas with area figures, (S6) finishing remarks. With this, every part of the interview could be assigned to a stage, the category is therefore exhaustive. With this structural coding RQ 1 could be addressed. By distinction between non-random patterns and more individual logics during problem solving RQ 2 could be addressed.

In the data analysis procedure, the first step was to define the stages of the interview: every part of the interview was assigned to a stage. In 20 out of 23 cases this could be done in the intended way concerning sequence and completeness. In one case the pupil found an acceptable solution at the S4 and does not need further aids. In this case, S5.1–5.3 were missing. In two cases the audios do not have a clear hint at which time the additional help is assigned and therefore cannot be coded. With this part of the coding manual, structural differentiation is done, the stages with more recall- and recognition-based aids meant to be differentiated. Interviewers were asked to clearly mention when they switch on a new aid.

Deductive coding was done for the three main categories "*problem solving phase*"; "*Instrumental Genesis*" (Rabardel, 2002) and "*problem solving strategies*". For the problem solving phase there were four subcategories defined, derived from the Problem Solving Cycle (Fernandez, 1994): understanding the problem; making a plan; carrying out the plan and looking back. For Instrumental Genesis (Rabardel, 2002) subcategories were deducted from an earlier study (Müller, 2023): instrumentation, instrumentalization, mediation, perceived competencies with DGS and influence of talking-out-loud. However, at this point the focus is on the main category of problem solving strategies.

Problem solving strategies were in the first step deducted logically by mathematically adequate strategies (ways to find a mathematically productive and adequate strategy which may lead to successful problem solving). We therefore defined the comparison of areas (arrangements and/ or figures), the comparison of edges, the comparison of slopes as such strategies. We also deducted mathematical strategies, which alone do not allow finding a coherent solution to the problem, as the comparison of simple length (no multiplication of height/ length, arrangements and/ or figures). Then a subcategory for other strategies were defined which surround all individual learner's strategies to solve the problem without a coherent mathematical background (e.g. the comparison of colors). This category then is subsequently analyzed in an inductive manner to identify problem individual reasonable or false problem solving strategies. The procedure therefore fulfils the demand of Schreier (2014, p. 3) in the context of content analysis to at least generate some of the categories in an inductive manner and to be open to surprising or not expected content (in our case, problem solving strategies). An overview of the definition of the subcategories is shown in Table 1 in the appendix. For problem solving strategies, S4 up to S5.3 were scanned according to utterances where pupils express some kind of comparison between the two arrangements or the combination of the two. Not coded within this category where utterances that are not related to the posed problem, manly containing the handling of DGS, simple descriptions of actions, simple descriptions of the problem itself, descriptions of arrangements (or its figures) without comparison or utterances of emotional kind (e.g. "I do not understand"). Because the main categories are not mutually exclusive, in contrast, they are meant to overlap, analysis could combine aspects of different main categories to hypothesize where subcategories cooccur or correlate and how they are interconnected.

In this article, we did two analyses: First (see Sect. 5.1), we analyzed a single case in depth concerning problem solving at different stages to verify the category system and to demonstrate the analyzing procedure. The case was chosen because all problem solving strategies were present. This is an exception; most pupils do not use all strategies. Second (see Sect. 5.2), 10 cases were coded by one rater concerning problem solving strategies. These 10 cases are analyzed in a quantitative manner by counting strategy use at the different stages. If a strategy is used one or more times throughout the stage it was counted as one.

5 First Results

5.1 Single Case Example

For this article we did the planned analyzing procedure for one learner, the subject L31. This example was selected because the boy showed all postulated comparison strategies and some characteristics of the interview situation that are, after first familiarization with data, exemplary for the whole sample. In the first problem solving stage (S4, no further aid is provided) L31 showed little understanding of the problem (see Fig. 1), exhibits less plans to address the problem and shows more on the spot approaches, in the sense that he begins repeatedly to count several lengths and heights of figures and arrangements. This procedure leads to a first but wrong conclusion: The missing square is due to differences in length of the two arrangements due to differences in order. He afterwards rationalizes his wrong hypothesis by adapting his count of length ("otherwise it wouldn't have fitted"):

"*I: Exactly, and can you explain why there wasn't just a little empty square there, in this second arrangement?*

P: I think it's because the shape has become a bit longer, lengthwise. (.) And then you'd have red and green, I mean no, because red wouldn't have fitted into the orange here, because then something might have been missing from the orange and it would have been a bit longer down here. But it would have fitted perfectly here and not here, and that's why you would have had to make a small gap here, because otherwise it wouldn't have fitted.

I: So, you're saying that the base length down here is shorter than this one?

S: Yes, because otherwise it wouldn't have fitted." (L31, positions 41–43, translation from German)

L31 here shows the strategy "comparison of lengths". As expected, this strategy alone does not result in a successful problem solving and, in this case, to a wrong conclusion or solution. He considered the problem to be solved, but by switching on the first digital aid (S5.1), he recognized by using the zoom-tool that the two arrangements differ by their slope (he uses the terms "steep" and "steeply"). This after first arguing further with his hypothesis of differences in order, he suddenly recognized that there are differences, he associated with knowledge or his concept of slopes. He expresses a thought close to an adequate argumentation but cannot point at it fully. The digital aid "zoom-tool" therefore enabled him by recognition to go in direction of an adequate solution strategy, the comparison of slopes:

"*I: What has changed?*

S: The shape, well, the second shape, it's slipped down a bit, because when the blue one was at the bottom, it was higher. And when the green one was at the top, the red one slipped down, the green one slipped down too, and the blue one went up. So it's sort of slipped down a bit. (..)

I: So you're saying that the height here is higher than here?

S: Well, no, the tops are the same height; it's just that when the line, (.) the straight line, goes down, it's no longer the same size, but the second figure goes down more steeply and the first figure is less steep. And because of that, the first figure is a bit bigger. (..)" (L31, positions 57–62, translation from German [tfG])

Interesting, L31 do use the digital aid "straight line", but he does not make a plan or had a coherent idea of how or where to use the straight line. With switching on the second digital aid (S5.2), he newly adapts his problem solving strategy, this time he seems to go a step back: due to the digital aid "auxiliary line" he now perceives the differences in the figures (convex and concave figures for each arrangement) but did not longer argue with slopes. The strategy "comparison of figures" is adopted. The digital aid seems to confuse L31, so he dropped the correct explanation. In this point L31 differs from other participants, in that for many learners the aid "auxiliary line" helps them the first time during the problem, to perceive differences in figures and develop a coherent problem understanding and to develop a plan and a problem solving strategy. However, L31 shows, as other learners, that problem solving in (geometric) problems is not a one-directed progression, but an iterative process, also with regression (see Sect. 2.2). Perceptual salient features here lead to recognition of fewer helpful problem solving strategies.

> "*S: Yes, on the longest one. (..) You can just see that they're not quite straight, a bit curved. [...]*
>
> *S: On the first one, it goes down and the black one goes up. And on the second one, the black one goes down and the first one goes up, so down.*
>
> *I: Mhm, you've observed that very, very well. (..) Can you explain that to yourself now? (..) Or... or...*
>
> *S: Um... Maybe... (11-second pause) No, not really.*
>
> *I: So you're saying that in this one, the line isn't quite straight; it's a bit lower than the correction line, and you're saying it's a bit above it?*
>
> *S: Mhm.*
>
> *I: Exactly that. Okay, and why is that? (..)*
>
> *S: Perhaps to make this figure a bit bigger.* " (L31, positions 93–98, tfG)

With the third digital aid (S5.3) there is possible a direct path to an adequate problem solving drawn: concrete numbers are provided that allow a comparison of areas. What L31 does, he switched to the strategy "comparison of areas". But in a not correct manner. He recognized the connection between the area under the auxiliary line (A = 0.5) and the area of the missing square (A = 1). What is missing is seeing the double arising of the area of 0.5 under and over the auxiliary line. With this digital aid, L31 changed into an adequate problem solving strategy, but in this case, he seems to incomplete recognize possible knowledge but only focuses on the visible salient features (the numbers 0.5 and 1).

> "*S: So the area of this triangle here, of this hidden one here, is 0.5, whereas the area of the square is 1. And that's why this one is bigger than that one. (..) Because theoretically... Because 1 is already more than 0.5. (...) But now it could also be the same, because there it's a bit less. And there it also has an area of 3 again.*
>
> *I: Exactly. (....)*
>
> *S: That's like a triangle too. (..) So theoretically, this one has a triangle and a square, so it's like 1.5. (.) And that one has 0.5. (...)*" (L31, positions 134–137, tfG)

Interestingly, L31 cannot come to a mathematically adequate problem solving despite he seems to be one a promising path.

5.2 Strategies According to Stages

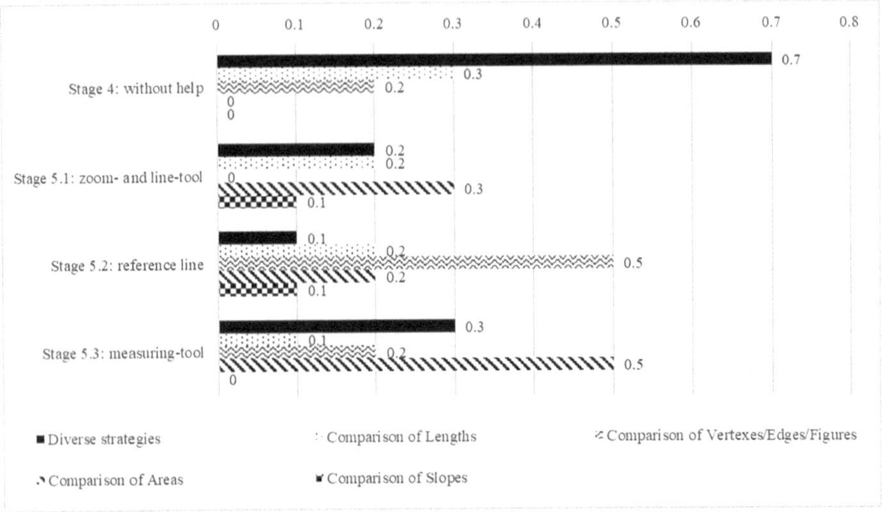

Fig. 3. Shown strategies according to stages in percent (n = 10).

Counting strategies according to stages show that in the first stage, where there were no explicit aids available, most pupils (7 cases) show no coherent strategies as predefined but argue in non-mathematical manner (diverse strategies). In this stage, lower strategies as comparison of lengths (mostly by counting, 3 cases) and the comparison of Figs. (2 cases) were partly present. With first aid, most students also used new problem solving strategies and more sophisticated strategies were used (comparing areas (3 cases) and slopes (1 case)). For 3 pupils the aid seems to be no resource for reorientating their thoughts, they remain by their arguing with diverse strategies. The reference line (S5.2) triggers mostly the strategy of comparing vertexes, edges and figures. Generally, the use of mathematically coherent strategies rises. The most strategies are found in the last stage, where comparison of areas became the most dominant strategy (6 cases). Overall, the strategy of comparing slopes is only used by 2 pupils during problem solving.

6 Discussion and Outlook

The criteria of the think-out-loud method were observed and implemented during the study (Konrad, 2010). The sample size should be increased to reliably assume empirical saturation regarding the described category system and code manual. The next step is reliability testing or improvement through data comparison with a second rater to make the category system reproducible. In terms of the diagnostic focus, it can be assumed, due to the in-depth follow-up questions, that the bias of overestimating the learners' knowledge is greater than the bias of underestimating it.

With respect to RQ1 Fig. 3 shows that all four described problem solving strategies are used by the 10 cases, which were examined by one rater so far (see Sect. 4.4). It is interesting that overall, just two students used the strategy of comparing slopes. Due to the age group (students on primary level) it must be considered that the concept of slope was not introduced to them yet, whereas the concepts of vertexes, edges, lengths and areas are familiar to them. Against this background it is understandable that comparing slopes is not the first choice of students to argue with.

First, with respect to RQ2 it can be stated students choose mathematically insufficient problem solving strategies, if there is no digital aid available (in 8 of 10 cases, see Fig. 3). Second, it can be noted that by introducing digital aids more sophisticated strategies (such as comparing vertexes, edges or figures and comparing areas) were chosen. Third, it is notable that different digital aids evoke different problem solving strategies. Especially the second digital aid evokes the problem solving strategy of comparing vertexes, edges or figures (in 5 of 10 cases, see Fig. 3): The auxiliary line gives a strong argument, that both arrangements cannot be assumed as triangles in total. Because of digital aid students are aware of this fact and investigate the type of figure by comparing vertexes or edges. Moreover, the third digital aid which highlights the additional areas in both arrangements evokes the problem solving strategy of comparing areas (in 6 of 10 cases, see Fig. 3): The highlighted areas in addition to the digital measuring tool of areas give a strong argument about the complementing figures and the relationship between the areas. Students get an idea of this argument and try to explain the difference between the two arrangements regarding the area.

The extent of the transfer of strategies in geometric problem solving with DGS becomes particularly visible in this context (Heinrich et al., 2015, p. 297). For a follow-up study, the systematic evaluation regarding the auxiliary line is relevant. It is already foreseeable that the second digital aid (auxiliary line) will set a decisive stimulus. It is therefore intended to test the following hypothesis quantitatively in a larger sample: Each of the three different digital aids (S4 - S5.3) evokes at least one of the four problem solving strategies (see Table 1 in the appendix).

In particular, the second digital aid (S5.2) evokes the more adequate problem solving strategies: comparison of slopes and comparison of figures. This allows for alignment with known strategies in the meaning of the recognition paradigm (here activation of adequate problem solving solution strategies). Moreover, it is important to compare the transfer potential with other problem-specific strategies in connection with further geometric problems (see Appendix) while using DGS (e.g. Müller, 2023; Müller, Weißing, & Lütscher, 2025).

Acknowledgments. We do say thank you to all 23 learners who participated in this study and in the enrichment programs of the University for Teacher Education of the Grisons. Moreover, we thank our student assistants who helped to collate all the data and double-check the transcriptions protocols.

Disclosure of Interests. The authors have no competing interests.

Appendix

As described in Sect. 6 further research will include a comparison of the transfer potential with other problem-specific strategies in connection with two geometric problems A and B, while using DGS. Whereas Problem A is described in Müller, Weißing, and Lütscher

(2025). Problem A shall be explained briefly: We are looking for the length of the shortest closed polygonal chain on the surface of a regular tetrahedron that intersects all four faces. Using the measurement-tool of a DGS gives a first argument that the length is double-size of one edge of the regular tetrahedron. In addition to determining the length, the question of the number of such polygonal chains is also of interest. A look at the surface net of the tetrahedron offers a strong argument for the infinitesimal many solutions of such shortest polynomial chains (Müller, 2024) (Fig. 4).

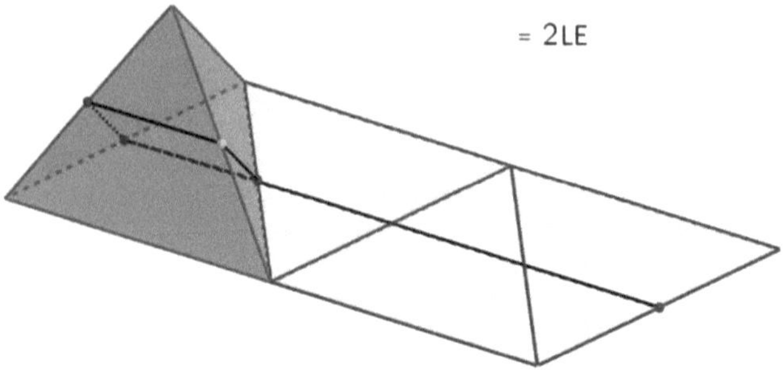

Fig. 4. Shortest distance on tetrahedron surface. Digitally measured path on 3D model and 2D surface net (Müller, 2024).

Table 1. Subcategories of "*problem solving strategies*" with definition/ exclusive definition.

Sub- category	Definition	Exclusive Definition
Comparison of Lengths	For both arrangements, line segments are described and identified. The lengths of comparable line segments are determined (digital or analogue) and compared. These may be heights, widths or edges. The focus remains on the consideration of line segments or pairs of line segments (one-dimensionality). From this, it can be indirectly concluded that the two arrangements do not represent triangles and must differ	There is no determination or description of line segments. No lengths of heights, widths or edges are compared

(*continued*)

Table 1. (*continued*)

Sub- category	Definition	Exclusive Definition
Comparison of Vertexes/ Edges/ Figures	The number of vertexes is determined for both arrangements. It is determined whether points are on edges, including auxiliary lines, (not a vertex of the composite figure) or not (a vertex of the composite figure). A description of the composite figure is provided. The pupils compare and describe figures in both arrangements (quadrilateral and irregular octagon) in terms of their vertexes or figures. One or two indented vertexes may be described (these are concave figures; missing parts of the arrangements)	The number of vertexes is not determined, nor is it checked whether points lie on edges (or auxiliary lines) or not. (especially when using the digital aid 'line')
Comparison of Areas	In both arrangements (digital or analogue), the widths and heights are determined ("measured") and multiplied (cf. Formula for area). Alternatively, in both arrangements, the areas are determined directly and compared (digital: measuring tool; analogue: counting squares). The conclusion may be drawn that the areas of the four shapes are equal but are distributed differently in the two arrangements. Two auxiliary triangles (area of 0.5) may be described	The (proportional) product of height and width is not determined. No invariance of the areas is formulated. No auxiliary triangles are described

(*continued*)

Table 1. *(continued)*

Sub- category	Definition	Exclusive Definition
Comparison of Slopes	In both arrangements, the widths and heights are determined ("measured") and compared in terms of their ratio (cf. Slope triangle). Alternatively, in both arrangements, the angles of inclination are determined and compared. The conclusion is drawn that the two arrangements differ because the slopes differ in this respect	Neither the ratio of height to width nor the angle of inclination is determined. Measurements of the height or length of arrangements are made but are not explicitly related to one another
Diverse strategies	Residual category for problem solving strategies that cannot be classified	

References

Baddeley, A.: The concept of episodic memory. Phil. Trans. R. Soc. Lond. Ser. B: Biol. Sci. **356**(1413), 1345–1350 (2001). https://doi.org/10.1098/rstb.2001.0957

Barzel, B., Büchter, A., Leuders, T.: Mathematik Methodik - Handbuch für die Sekundarstufe I und II. Cornelsen Scriptor, Berlin (2007)

Börnert, M., Wilbert, J.: Thinking-aloud protocols of piagetian tasks: insights into problem-solving processes of primary school students. Insights Learn. Disabil. **12**(1), 19–34 (2015)

Elward, R., Limond, J., Chareyron, L.J., Ethapemi, J., Vargha-Khadem, F.: Using recognition testing to support semantic learning in developmental amnesia. Neuropsychol. Rehabil. **34**(8), 1141–1160 (2024). https://doi.org/10.1080/09602011.2023.2275825

Ericsson, K.A., Simon, H.A.: How to study thinking in everyday life: contrasting think-aloud protocols with descriptions and explanations of thinking. Mind Cult. Act. **5**(3), 178–186 (1998). https://doi.org/10.1207/s15327884mca0503_3

Fahy, J.K.: Language and executive functions: self-talk for self-regulation. Perspect. Lang. Learn. Educ. **21**(2), 61–71 (2014). https://doi.org/10.1044/lle21.2.61

Fernandez, M.L., Hadaway, N., Wilson, J.W.: Problem solving: managing it all. Math. Teach. **87**(3), 195–199 (1994). https://doi.org/10.5951/MT.87.3.0195

Funke, J., Spering, M.: Methoden der Denk- und Problemlöseforschung. In: Funke, J. (eds.) Enzyklopädie der Psychologie. Denken und Problemlösen, pp. 647–744. Göttingen: Hogrefe (2006)

Haist, F., Squire, L.R., Shimamura, A.P.: On the relationship between recall and recognition memory. J. Exp. Psychol. **18**(4), 691–702 (1992)

Hasselhorn, M., Schneider, W.: Handbuch der Entwicklungspsychologie. Hogrefe, Göttingen (2007)

Heinrich, F., Bruder, R., Bauer, C.: Problemlösen lernen. In: Bruder, R., Hefendehl-Hebeker, L., Schmidt-Thieme, B., Weigand, H.-G. (eds.) Handbuch der Mathematikdidaktik, pp. 279–301. Springer Spektrum, Berlin (2015)

Imhof, A.L., Kübler, M.: How to reliably diagnose children's concepts in learning science? Using the water cycle as an example. Interdisc. J. Environ. Sci. Educ. **21**(2), e2509 (2025). https://doi.org/10.29333/ijese/15960

Konrad, K.: Lautes Denken. In: Mey, G., Mruck, K. (eds.) Handbuch Qualitative Forschung in der Psychologie, pp. 476–490. VS Verlag für Sozialwissenschaften, Wiesbaden (2010)

Müller, M.: Lehren und Lernen mit digitalen Mathematikwerkzeugen. Berücksichtigung von institutionellen, individuellen und sprachlich-kulturellen Bezügen der instrumentalen Genese. Springer, Wiesbaden (2023).https://doi.org/10.1007/978-3-658-41115-2

Müller, M.: Entdecken durch Problemlösen – eine qualitative Analyse zweier digitaler Lernumgebungen. Der Mathematikunterricht **70**(4), 24–34 (2024)

Müller, M., Weißing, B., Lütscher, P.: Solid geometry modeling: 3D printing is not always the best option. In Fernau, H., Schwank, I., Staub, J. (eds.) Creative Mathematical Sciences Communication. CMSC 2024. Lecture Notes in Computer Science, vol. 15229. Springer, Cham (2025). https://doi.org/10.1007/978-3-031-73257-7_7

Müller, M., Imhof, A.: Vorschlag zum methodischen Design einer Untersuchung zur Arbeit mit digitalen Mathematikwerkzeugen mittels der Methode des lauten Denkens. In: Ambrus, G., Sjuts, J., Szűcs, K., Vásárhelyi, É. (eds.) Schlüssel zum Erfolg: Kognitive und metakognitive Prozesse beim Verstehen von Mathematik, pp. 187–206. WTM, Münster (2025). https://doi.org/10.37626/GA9783959873505.0

Ostermann, A., Lindmeier, A., Härtig, H., Kampschulte, L., Ropohl, M., Schwanewedel, J.: Mathematikspezifische Medien nutzen: Was macht den Unterschied – Lehrkraft, Schulkultur oder Technik? DDS – Die Deutsche Schule **2021**(2), 199–217 (2021). https://doi.org/10.31244/dds.2021.02.07

Pólya, G.: How to solve it: a new aspect of mathematical method/G. Polya: a new aspect of mathematical method. Princton science library printing. Princeton Univ. Press (1954)

Raaijmakers, J.G.W., Shiffrin, R.M.: Models for recall and recognition. Ann. Rev. Psychol. **43**(1), 205–234 (1992). https://doi.org/10.1146/annurev.ps.43.020192.001225

Rabardel, P.: People and technology. université paris 8, pp.188. ⟨hal-01020705⟩ (2002)

Rhodes, S., Greene, N.R., Naveh-Benjamin, M.: Age-related differences in recall and recognition: a meta-analysis. Psychon. Bull. Rev. **26**(5), 1529–1547 (2019). https://doi.org/10.3758/s13423-019-01649-y

Rott, B., Specht, B., Knipping, C.: A descriptive phase model of problem-solving processes. ZDM-Math. Educ. **53**, 737–752 (2021). https://doi.org/10.1007/s11858-021-01244-3

Schreier, M.: Varianten qualitativer Inhaltsanalyse: Ein Wegweiser im Dickicht der Begrifflichkeiten. Forum qualitative Sozialforschung, 15(1), 18 (2014)

Tversky, B.: Encoding processes in recognition and recall. Cogn. Psychol. **5**, 275–287 (1973)

Vygotsky, L.S.: Mind in Society: Development of Higher Psychological Processes. Harvard University Press, Boston (1978). https://doi.org/10.2307/j.ctvjf9vz4

Developing Higher-Order Reasoning

Mapping Abstract Reasoning: A Freinet-Inspired Approach to Creative Logic Communication in Secondary Schools

Cosimo Perini Brogi[✉]

IMT School for Advanced Studies Lucca, 55100 Lucca, Tuscany, Italy
cosimo.perinibrogi@imtlucca.it

Abstract. The teaching of the mathematical foundations of computer science – in particular, mathematical logic – is frequently hampered by a high degree of abstraction, leading to disengagement and significant barriers to student comprehension. Traditional pedagogical approaches, often centred on passive instruction and formal definition, struggle to convey the dynamic applicability of these concepts. This paper explores a novel application of Célestin and Élise Freinet's pedagogy – originally developed for primary education – to the context of secondary school, for classes in mathematics, and, possibly, philosophy. We argue that Freinetian principles, such as experimental trial-and-error (tâtonnement expérimental), cooperative learning, and the production of artefacts, offer a robust remedy to the "theory-practice gap" in formal sciences, and also cultivate creativity along with essential transversal skills. We present a concrete communication proposal anchored in an integrated didactic unit, we named "The Logic Treasure Map": the creation of a simple interactive experience serving as a narrative thread, grounding abstract concepts in tangible design. As a conceptual contribution, it offers a theoretically grounded framework for practitioners, paving the way for future empirical evaluation.

Keywords: Freinet Pedagogy · Computer Science Unplugged · Mathematical Logic · Active Learning · Cooperative Learning · Creative Teaching Methods

1 Introduction

Conventional pedagogical approaches – predominantly teacher-led instruction and the mechanical resolution of standardised exercises – frequently fail to convey the sense of discovery, intellectual play, and dynamic applicability essential for motivating students in mathematical logic and the foundations of computer science [14,21,32].

Rather than fostering a deep, transferable understanding of formal concepts, such methodologies inadvertently create epistemological obstacles [7]. In logic,

M. Löffler et al. (Eds.): CMSC 2026, LNCS 16951, pp. 241–256, 2027.
https://doi.org/10.1007/978-3-032-38887-2_18

this manifests as a profound "syntax-semantics gap" [13]: students perceive formal proofs and operations as the arbitrary manipulation of meaningless symbols, instead of systematic tools for rigorous reasoning [4]. The traditional "top-down" delivery of definitions and axioms (or formal rules) imposes a finished mathematical product, bypassing the vital heuristic process of genuine inquiry [39] and often exacerbating the anxiety associated with formal disciplines [28,34,40].

To dismantle these barriers, this paper proposes a recontextualised application of Freinet pedagogy, an educational framework developed by the French educators Célestin and Élise Freinet in the early twentieth century [15,17]. Although originally conceptualised for primary education, the foundational pillars of the Freinet method – student-centred exploration, experiential learning, cooperative organisation, and free expression – translate naturally to the formal sciences when rigorously adapted [37,44].

Contributions. The primary contribution of this work is the detailed presentation and theoretical justification of a standalone teaching unit titled *The Logic Treasure Map*.

Designed for upper-secondary mathematics or philosophy, the unit reconciles formal rigour with logic communication aimed to active student engagement. It leverages the narrative allure of a treasure hunt to translate abstract propositional logic into a tangible, rule-based communicative medium, resonating with Papert's constructionism and the creation of community-shared artefacts [22,35].

Through the cooperative design of a choice-driven map, students engage in *tâtonnement expérimental* (experimental trial-and-error), collaboratively negotiating logical syntax to resolve concrete structural problems. This process respects the psychological laws of learning [16] and mirrors the iterative debugging inherent to computer science, repositioning failure as a vital learning step rather than a punitive error [27,41].

Furthermore, by situating this negotiation within a community of practice *and* learning [20,29], the social validation of knowledge through peer interaction deepens understanding [24,36,45] – in this case, of the rudiments of mathematical logic. By inviting students to explore logical structures experimentally – manipulating propositions, constructing scenarios, and observing consequences – teachers can cultivate an intuitive, semantic grasp of operations prior to formalisation.[1] Consequently, the educator acts not as a distant authority but as a facilitator of inquiry [30,33].

While the design is informed by pedagogical practice, this paper deliberately presents a conceptual framework rather than an empirical study; its goal is to provide a rigorous, replicable blueprint for practitioners. Through this example, distilled from some preliminary experiments in self-organised learning environments, we posit that the objective of creatively communicating mathematical logic is to cultivate autonomous, computationally competent learners who can

[1] This sequencing addresses documented developmental challenges identified, across several education levels, in proof construction and mathematical proficiency [2,12, 25].

participate creatively in an evolving scientific landscape – not merely to enable students to pass examinations.

Paper Structure. The remaining parts of the paper are structured as follows. Section 2 presents the core didactic proposal, "The Logic Treasure Map", outlining its modular design, the five-phase implementation process, and strategies for interdisciplinary adaptation and formative evaluation. Section 3 critically examines the advantages and logistical constraints of this communication method, addressing potential epistemological tensions and identifying possible protocols for future empirical verification. Section 4 concludes by affirming the value of democratic, constructionist approaches in revitalising the communication of formal sciences. Finally, Appendix A delineates the theoretical underpinnings of the Freinets' approach, explicating the principles of tâtonnement expérimental, cooperative social validation, and artefact production, whilst contextualising the systemic barriers within secondary education.

2 The Logic Treasure Map

In this section, we present the teaching unit *The Logic Treasure Map* as a direct Freinetian response to the pedagogical challenges outlined above. The unit replaces rigid top-down instruction with a progressive, bottom-up experiential design and communication of the rudiments of mathematical logic.

Requiring approximately 6 to 8 h of dedicated classroom time, the unit is highly modular and readily adaptable to existing curricular constraints.

2.1 Global Design and Pre-Requisites

The core objective is for students, working in small *cooperative* teams of three to four, to design a physical "treasure map" governed entirely by strict, explicit logical conditions. Functionally, the map operates as a complex decision tree or state-transition diagram embedded within an engaging narrative shell.[2] To navigate the map successfully and locate the "treasure", a player must satisfy various interlocking propositional requirements dictated by environmental obstacles, items, and choices.

The physical materials are deliberately minimal, drawing inspiration from the *Computer Science Unplugged* movement [8]:[3] large sheets of poster paper, coloured markers, and a structural "set of elements" provided by the teacher. By requiring no prior familiarity with computer programming or formal mathematical notation, the activity lowers initial barriers to entry and establishes an equitable baseline for all students.

[2] See Fig. 1 below for an abstract and rudimental rendering of such a map.

[3] The Computer Science Unplugged movement is highly congruent with the Freinets' viewpoint on proposing relevant experiential activities, as witnessed by the movement's official website: https://www.csunplugged.org/en/.

2.2 Phase-by-Phase Implementation

Phase 1: Narrative Trigger and Activation (approx. 45 min)
The unit opens with a collective brainstorming session. The instructor presents a classic illustrative map featuring traditional adventure tropes (e.g., rope bridges, volcanic chasms, hidden chests). Rather than lecturing on logic, the instructor introduces a novel narrative constraint:

> Imagine this map is alive, and its pathways open or slam shut based on precise, absolute laws. How can we write the laws of this world so that only a truly logical explorer can survive?

Through structured, interactive dialogue,[4] the instructor elicits natural-language equivalents of logical connectives. For instance:

> If a path requires a rope bridge to be intact *and* a secret word to be spoken, what happens if you only have the word but the bridge is broken?

This immediately establishes the semantic truth-value of a conjunction without invoking formal notation. The same inductive questioning is deployed for disjunction, negation, and the conditional.

By externalising these rules collectively, students engage in a process of cognitive scaffolding [6], collaboratively building a shared vocabulary of conditions, states, and consequences. This structured dialogue provides the necessary conceptual toolkit for Phase 2, equipping students to begin negotiating, designing, and structuring their own complex logical environments.

Phase 2: Cooperative Design of the Map (approx. 90 min). Students divide into their established cooperative teams. Each group receives a large poster sheet and a set of mandatory structural components that must be woven into their design. To preclude trivial designs, the teacher mandates specific constraints, such as:

- A definitive starting node and a final treasure node.
- At least two distinct environmental hazards or obstacles (e.g., a dragon's lair, an impassable river, a shifting labyrinth).
- At least one ambient environmental constraint (e.g., a dark cavern) that functions as a convergent choke point.
- Utility items or internal states acquirable along the way (e.g., a magic sword, a rowboat, a torch).

During this phase, classroom dynamics mirror a Freinet workshop. Students engage in lively, argumentative design, negotiating rules and mapping them visually. A student might propose:

> Let's make it so you can cross the river if you have a boat or if you know how to swim.

[4] Such a dialogue may be rooted in the principles of, e.g., *guided reinvention* [19] and *dialogic teaching* [3].

Another student immediately acts as a logic verifier:

> But wait! If you choose to swim, you cannot carry the magic sword because it is too heavy, so you must drop it!

Here, *tâtonnement expérimental* is vividly operational: the team discovers that an early narrative choice has cascading structural consequences for downstream logical expressions. The teacher circulates, observing, offering reflective prompts, and stimulating discussion on rule coherence.

The emergence of diverse solutions is treated as an asset; it may even prompt the instructor to *informally* discuss advanced meta-theoretical issues – e.g., the inter-definability of Boolean connectives, or extensions to modal logics.

Phase 3: Formalisation and Documentation (approx. 60 min). Once the visual design is complete, the instructor introduces the syntactic shift. Groups are informed that their maps cannot remain mere intuitive illustrations; they must be accompanied by an official, highly precise "Rule Manual" or "Logic Specification Sheet". To support this translation, the instructor may distribute a structural template featuring a variable dictionary (mapping narrative items to propositional letters) and a rule inventory table, depending on the needs of the participants of the workshop.

The instructor then demonstrates how to convert natural-language sentences into symbolic propositional variables and connectives:

> Let S represent 'Player has Magic Sword', D represent 'Player has Shield', and G represent 'Dragon Defeated'. The rule 'The dragon is defeated if and only if the player possesses both the magic sword and the shield' is formalised as $G \leftrightarrow (S \wedge D)$.

Crucially, the formalisation phase remains a cooperative, dialogic endeavour. Students are encouraged to negotiate the boundaries of their definitions, debating questions about the symbolic rendering of natural-language sentences. Formalisation is not treated as a unidirectional translation of a frozen design; rather, it acts as a logical stress-test within the practice of *tâtonnement expérimental.* As students attempt to symbolise their maps, they may frequently uncover structural flaws, unhandled edge cases, or unintended paradoxes in their Phase 2 designs. Iterative refinement of the physical map is thus not only permitted, but actively encouraged, with the act of writing formal notation driving the iterative debugging of the map's physical architecture.

The abstract architecture of a simplified map (Fig. 1) illustrates this translation, further extended into a basic narrative.[5] Notably, the Cavern block introduces an ambient environmental constraint: $\neg T \rightarrow$ *Blocked.* Pedagogically, this

[5] Naturally, initial student-drawn maps will be spatially complex and rich in aesthetic detail, bearing little visual resemblance to a formal state-transition diagram. Accompanying the formalisation process thus requires guiding students to abstract their drawings, depending on the capacities and needs of each students – see Sect. 2.3 below.

choke point is vital. Whether a player arrives via the Dragon ($S \wedge D$) or the River ($B \vee W$), they encounter a new, independent state variable (T). This enforces a "logical principle" of convergent path validation: past successes do not guarantee future progression if the requisite propositional variables are absent.

The teacher's public interactions during this phase guarantee that the documentation is clear and comprehensible even to those who did not design a specific map, whilst conveying the operational meaning of the connectives to the entire class.

This phase requires students to traverse the syntax-semantics gap deliberately. Formal notation is no longer an arbitrary imposition from an external authority; it is revealed as a compact, unambiguous language essential for codifying the mechanics of an environment they have personally authored.

Phase 4: The Map Market and Cross-Verification (approx. 60 min). Having completed their documentation, groups participate in the "Market of Maps", activating the Freinet principles of communication and social validation.

Groups display their physical maps alongside their formal Rule Manuals. Cooperatives then rotate systematically around the classroom, acting as external quality-assurance teams or "logic debuggers".

Upon receiving another team's map, a cooperative must "test" its structural integrity by simulating different play runs and boundary cases:

> What happens if a player arrives at the dragon with the sword but without the shield? Are there dead ends or logical paradoxes where a player satisfies all conditions but is blocked by a structural contradiction?

If an inconsistency is discovered, the testing team must document it as a precise "bug report", detailing why the formal propositional formula fails to align with the designers' visual or narrative intentions.[6]

This cross-verification reinforces logical mastery; students solidify their understanding not merely by writing their own rules, but by critically testing their own understanding of the formal frameworks engineered by their peers [45].

Phase 5: Collective Synthesis and Debriefing (approx. 45 min). The unit concludes with a teacher-facilitated plenary. Each group receives its peer-reviewed informal feedback and is granted a (brief) window to re-discuss, modify and correct any structural flaws. The instructor selects illuminating problems that emerged across maps for collective discussion:

> Team B discovered a loop where a player could infinitely generate keys without encountering an obstacle. How can we fix this rule using a negation or a conditional statement?

[6] Notice that no mark is assigned by the groups: the whole intervention is aimed to collaborative learning, communication and development, so that it is hardly comparable to a competitive gamification exercise.

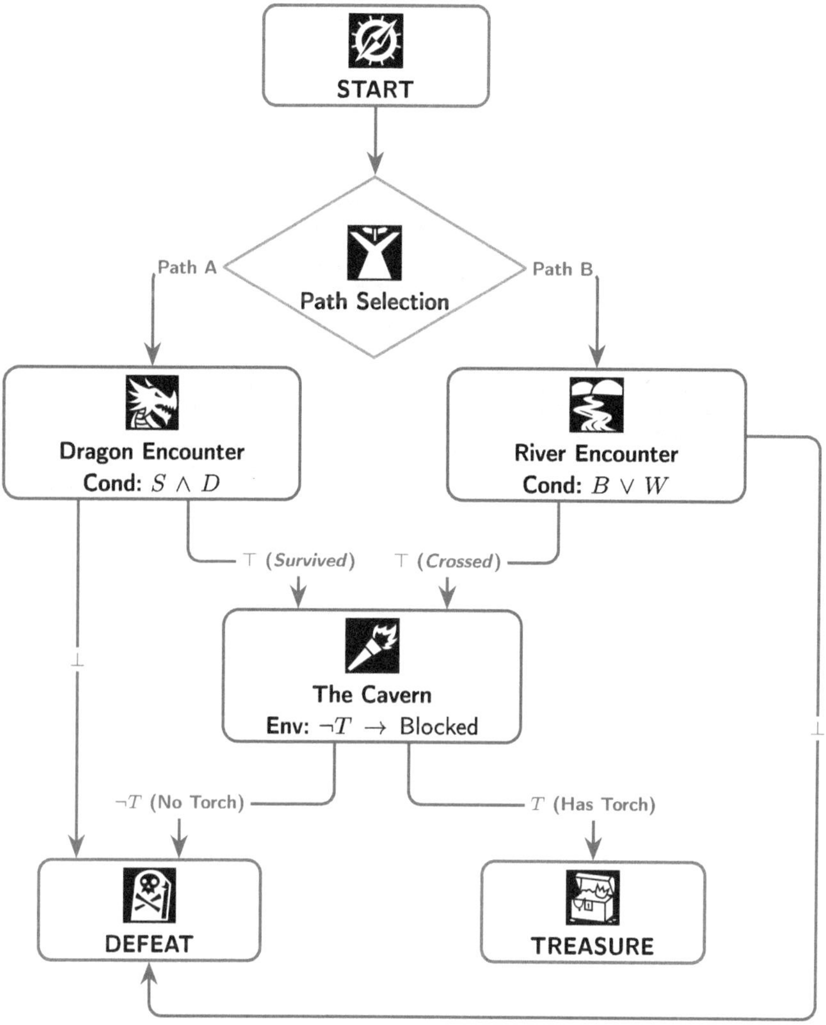

Fig. 1. Architectural logic flow of a simplified *Logic Treasure Map*. Variables represent player states: S (Magic Sword), D (Shield), B (Boat), W (Swimming Ability), T (Torch). State transitions are mediated by logical truth ($\top$) and falsity ($\perp$).

Finally, the completed maps and validated logical frameworks are displayed prominently in the classroom or published online, forming a permanent library of student-generated artefacts.

This phase allows for synthesis, the clarification of residual doubts, and the valorisation of diverse approaches. The teacher may pose reflective questions: "Which logical rule was the most difficult to design and why?", "Did you find different ways to utilise the same connective?", or "What did you learn by verifying your peers' work?".

2.3 Differentiated Adaptations for Educational Levels and Curricula

The structural flexibility of *The Logic Treasure Map* allows calibration across developmental stages and seamless integration into varying secondary-level curricula:

- Within a secondary mathematics curriculum, the unit serves as a concrete anchor for formal set theory, Boolean algebra, and early foundational mathematics.
- In secondary philosophy courses, the unit can be stripped of its mathematical formalism and deployed to explore semantic validity, the structure of fallacies, and deductive consistency.

Table 1 details some of these possible interdisciplinary adaptations.

Table 1. Interdisciplinary implementation matrix of The Logic Treasure Map.

Mathematics Focus	Philosophy Focus
Set Theory & Isomorphisms	Structural Validity vs. Soundness
Boolean Truth Tables	Argumentation & Rhetorical Paths
State-Transition Diagrams	Analysis of Fallacies
Algorithmic Pseudocode	Modelling Ethical/Existential Choices

Complementarily, to implement the Freinet principle of individualisation, the complexity of the design challenge should remain dynamically responsive to the student's zone of proximal development [46].

For groups requiring additional cognitive scaffolding, the activity can restrict its scope strictly to basic conjunctions ($\wedge$) and disjunctions ($\vee$). Consequently, in the formalisation/abstraction phase, the instructor may provide structural templates or guide the abstraction process directly, restricting the formal scope according to the operators at hand.

Conversely, advanced students can be tasked with independently abstracting their spatial maps into formal architectural flows. Furthermore, the framework scales quite naturally to more expressive languages, such as first-order logic. Instead of relying solely on propositional atoms, students can introduce predicates, variables, and quantifiers to govern refined map interactions.[7] By grounding abstract quantifiers and variables within concrete, spatially organised game mechanics, the map serves as an accessible sandbox for experimenting with relations and scope, as well as quantification domains. For adding further linguistic extensions, the instructor may introduce nested modal constraints,

[7] An existential constraint could dictate that a magical gate opens only if there *exists* a specific item in the player's inventory. Similarly, a universal constraint might require that *all* guardians along a designated path are pacified before traversal is permitted.

free-logic quantification, or require the implementation of non-standard operators like exclusive-OR, whilst anchoring them to concrete examples to manage intrinsic cognitive load [42]. For participants particularly revealing mathematically minded, real techniques for proof development could be introduced by the instructor.

Similarly, when participants show natural inclinations to critical thinking and analytical discussion of arguments, the instructor may introduce informal topics in philosophical logic, being inspired by the examples presented in more advanced textbooks. For instance, to explore alternatives to the material conditional, students might design a path where an implication only holds if the antecedent is strictly *relevant* to the consequent. They would quickly discover through play-testing that a rule like "If the player wears a red hat, then the dragon is defeated" fails to open the path, as the antecedent lacks causal or logical relevance, thereby exposing the paradoxes of material implication through spatial gameplay. Similarly, to move beyond two-valued reasoning, a map could introduce a third truth value for ambiguous or paradoxical states. A player might encounter a twilight zone where possessing a torch does not simply mean $T = \top$ or $T = \bot$, but rather an intermediate state. To traverse it, students could apply a many-valued or paraconsistent logic, allowing a player to hold contradictory states. By keeping the map's narrative and spatial mechanics intact but altering the underlying truth-conditions, the unit may thus naturally bridge formal game design with philosophical logic and non-classical reasoning.

2.4 Evaluation Mechanics

Aligned with the democratic tenets of Freinet pedagogy, the final evaluation synthesises three distinct feedback loops:[8]

1. **Self-Assessment:** At the conclusion of Phase 5, individual students complete a reflective logbook entry evaluating their personal growth, their comfort with logical syntax, and their specific contributions to the cooperative's output.
2. **Peer Review within the Group:** Team members informally evaluate each other's cooperative reliability, assessing variables such as communicative openness, willingness to share workloads, and supportiveness during design gridlocks. Anonymity is relevant here to enhance the collaborative nature of this phase and to prevent potential tensions among team members.
3. **Teacher Formative Observation:** The instructor acts as an ethnographer and technical facilitator throughout the unit, assessing group dynamics and mapping progress incrementally rather than relying solely on the final poster output.

Crucially, pedagogical transparency dictates that students are made fully aware of these evaluative criteria prior to commencing the design tasks, enabling

[8] These loops are intentionally designed to deactivate potential competitive dynamics that may emerge during the evolution and evaluation phases of the unit.

them to consciously direct their collaborative efforts. Within this framework, assessment is reframed as an intrinsic component of the learning trajectory – a dynamic instrument for cognitive and cooperative growth [20]. The goal is to significantly mitigate general performance anxiety, encouraging learners to shift their focus from static grades toward continuous, iterative improvement [5].

Evaluation is thus conceptualised as an ongoing, multidimensional, and formative process embedded directly within the collaborative learning activity. It considers not only technical accuracy but also problem-solving persistence, documentation quality, and cooperative engagement [9].

3 Critical Discussion

The primary advantage of *The Logic Treasure Map* is its immediate mitigation of "abstraction paralysis" in the mathematical sciences [23]. The non-competitive, cooperative matrix facilitates that less confident students are not left isolated; they are supported by a peer group that models problem-solving strategies in real time and, crucially, in a creative manner. Concurrently, the unit cultivates essential transversal skills: collaborative negotiation, precise technical documentation, resilience to peer critique, and systematic debugging.

Nonetheless, the principal concrete obstacle to implementation is logistical, and lies in the rigid, fragmented timetabling of traditional secondary schools. Splitting a 6-to-8-h experiential unit across disconnected 45-minute periods separated by several days can severely disrupt group momentum and dilute narrative engagement.

Furthermore, the translation of formal logic into a narrative medium may introduce an inherent epistemological tension. For instance, by embedding logical conditions within a sequential, spatial narrative – the traversal of a map – there is a risk that students might conflate logical implication with temporal causation. To counteract this specific misconception, the Phase 5 debrief may be leveraged to explicitly untangle sequential gameplay from truth-functional and operational semantics of the connectives. In case of such a misconception, the instructor may guide students to recognise that while a treasure map is traversed over time, propositional formulas evaluate states instantaneously. For instance, targeted discussion prompts can be deployed during the plenary: "If possessing the torch (T) implies you can enter the cavern (C), formalised as $T \to C$, does this mean the torch *causes* the cavern to open? Or does it simply mean that at the precise moment you stand at the entrance, the condition T must be satisfied?" By interrogating the difference between causal triggers and logical preconditions, students are guided to perceive implication as a purely functional – static, in the realm of classical logic, indeed – and abstract relation between statements, rather than a mechanical connection between facts.

Additionally, the allure of the creative shell may occasionally overshadow the formal architecture it is meant to instantiate, reducing the activity to mere aesthetic distraction rather than a genuine cognitive scaffold. Guaranteeing that the creative medium serves as a rigorous vehicle for formalisation, rather than a superficial veneer, demands thus careful pedagogical orchestration, significant pedagogical flexibility, and a high level of Pedagogical Content Knowledge (PCK) [38]. Instructors accustomed to absolute instructional control and linear, predetermined delivery may struggle with the decentralised, dynamic nature of a Freinet workshop.[9]

3.1 Empirical Verification Protocols

It is important to explicitly note that the current contribution is conceptual in nature; no formal empirical data or student-generated artefacts are presented in this work. The design of the unit and its constraints has been informed by foundational pedagogical theories and heuristic trials, but its systemic efficacy remains to be empirically measured according to well-established methodologies. To measure the long-term cognitive efficacy of this Freinet-inspired communicative intervention compared to traditional ex cathedra logic lecturing, future work requires a rigorous, mixed-methods empirical study. Controlled cohorts could be assessed across metrics tailored to the specific mechanisms of the intervention:

- Quantitative tracking of the "syntax-semantics gap", measuring students' capacity to fluidly translate between natural-language narratives and formal symbolic notation, alongside long-term conceptual retention.
- Qualitative ethnographic mapping of *tâtonnement* dynamics, observing how students utilise failure and iteration (debugging) to refine their logical models, and measuring shifts in mathematics anxiety indices.

Such empirical validation would expand the systemic reach and scalability of the *Logic Treasure Map* paradigm anticipated here. Ultimately, data could be compiled into an open-access repository, providing secondary instructors with an empirical baseline for validating Freinet-inspired, active-learning strategies within formal science syllabi.

4 Conclusion

The instruction of mathematical logic in secondary education may stand at a critical impasse. Continuing to rely on traditional, passive symbol manipulation risks alienating generations of students, obscuring the profound intellectual beauty of the formal sciences behind a wall of premature abstraction.

We have presented *The Logic Treasure Map* as a vital, humanising alternative that integrates the progressive insights of Célestin and Élise Freinet [1,15]. By

[9] The teacher must tolerate productive noise, manage interpersonal conflicts, and diagnose idiosyncratic logical errors on the fly without a standardised answer key. Should group dynamics fail – for instance, if a dominant student monopolises the design process – the cooperative structure can degrade into an unequal workload, reinforcing the very inequities the method seeks to dismantle. Such dynamics require careful and continuous facilitation [10].

anchoring formal propositional syntax within a tangible, cooperative narrative design space, the unit seeks to transform the classroom from a site of passive consumption into a vibrant, democratic laboratory of collective discovery, mathematical creativity, and free scientific communication [16,18].

Through *tâtonnement expérimental* (experimental trial-and-error), peer validation, and the production of authentic artefacts, students may actively dismantle the syntax-semantics gap. They discover that mathematical rigour is not an arbitrary institutional hurdle, but an empowering, expressive language for creative construction and rigorous problem-solving.

Ultimately, this proposal advances a precise philosophical statement regarding the communication of the mathematical sciences and cognitive growth: it is by daring to experiment, by dissolving competitive structures, and by trusting the autonomous agency of the classroom cooperative that we can revitalise the dissemination of logic and the mathematical foundations of computer science for the twenty-first century, potentially extending beyond traditional school and educational environments.

Acknowledgments. The author acknowledges the support of the European Mathematical Society, the INdAM-GNSAGA group, and The Proof Society.

Sincere gratitude goes to three anonymous reviewers for their time and the guidance they provided in refining this work.

Many thanks to the kindred spirits who drifted into the orbit of *La Rosa de Foc*. Their curiosity helped germinate the idea of employing the Freinets' approach to convey the innate creativity of logic to a wider audience.

The icons used in Fig. 1 were created by Delapouite, Skoll, and Lorc; they are obtained from https://game-icons.net/ under a CC BY 3.0 license.

Disclosure of Interests. The author has no competing interests to declare that are relevant to the content of this article.

A Theoretical Foundations

The pedagogical framework of Célestin and Élise Freinet is fundamentally rooted in the learner's intrinsic capacity to construct knowledge through direct environmental interaction, expressive production, and structured socialisation [15,17]. The Freinets maintained that education must be an active laboratory for life rather than a passive receptacle for detached facts, thereby cultivating responsible, autonomous agents of their own cognitive and communal development [16,18]. When mapped onto the epistemological demands of mathematical logic, core Freinetian principles emerge as deeply transformative, potentially.

Tâtonnement Expérimental (Experimental Trial-and-Error). Central to the Freinet method is the principle of *tâtonnement expérimental* (experimental trial-and-error). Traditional mathematics instruction often pathologises error, treating mistakes as definitive failures of comprehension; Freinet pedagogy, conversely, conceptualises error as the primary mechanism of epistemic advancement [27].

In logic, students cannot truly grasp the systemic stability of a truth table or deductive rule-set through memorisation alone. They must actively "tinker" with logical conditions, generate hypotheses, construct faulty arguments, encounter systemic contradictions, and refine their models iteratively [41]. By inviting students to manipulate logical rules within a *concrete, contextualised narrative*, tâtonnement facilitates a robust semantic intuition prior to the introduction of rigid, formal syntax.

The Classroom Cooperative and Social Validation. Freinet pedagogy replaces the hyper-individualistic, competitive matrix of the traditional classroom with the structure of the cooperative workshop. Learning is recognised as an intrinsically social endeavour, wherein each student contributes their abilities and learns from the diverse temperamental and cultural perspectives of others [29]. Applied to mathematical foundations, the classroom becomes a collective research community, driven by individual or team inquiries into themes relevant to the broader scientific community [26].

Within *non-competitive* group structures, students are required to externalise their reasoning, defend their logical designs, and collectively negotiate the validity of their assertions. This cooperative negotiation fundamentally shifts the source of mathematical validation: truth is no longer dogmatically conferred from above by the omniscient instructor; instead, it emerges democratically through logical consensus and rigorous peer verification.

Production of Artefacts and Free Expression. Freinet pedagogy encourages free expression through techniques such as *texte libre* (free text), school newspapers, and inter-school correspondence [17]. Recontextualised for formal sciences, learning and communicating logic becomes an act of constructing functional logical systems. Free expression is achieved when students possess the autonomy to design their own narratives, rule-sets, and pathways.

By writing documentation, creating visual schemas, or designing logical problems for peers, students cease to be consumers of a finished discipline; they become authors of intellectual discoveries, imbuing the learning process with authentic purpose and social utility. This dynamic consolidates learning and develops essential skills in scientific communication and technical language mastery [31], aligning closely with Papert's constructionism, which posits that learning occurs most felicitously when the learner constructs a meaningful public artefact [22].

Contextualising Challenges in Secondary School Logic Education. To appreciate the relevance of a Freinet-inspired intervention, one must diagnose the pedagogical failures of secondary-level logic instruction. Mathematical logic, in particular, often occupies a precarious position in the curriculum: frequently isolated as a brief introductory chapter in mathematics textbooks, or treated as a dry historical footnote in philosophy courses.

The primary failure of traditional instruction lies, in any case, in its radical commitment to a top-down trajectory. Concepts are presented purely theoretically – definitions, theorems, and proofs "imposed from above" – without sufficient anchoring to concrete instances, practical applications, or heuristic analogies – which are meaningful within the precise education community – that might later be refined into formal abstractions [43].

Coupled with an individualistic framework, this approach reinforces exclusionary myths about innate ability [11] and exacerbates mathematics anxiety. When confronting complex abstractions in isolation, a student's fear of failure frequently inhibits cognitive exploration; the fear of appearing incompetent induces risk aversion, prompting passive compliance or disengagement rather than the testing of non-standard hypotheses [28,34,40].

Conversely, the social validation of knowledge – where truth emerges from non-competitive peer interaction rather than solely from the instructor – provides a collaborative scaffold and an experimental, low-stakes environment. It is within this environment that true mathematical maturity develops [45] and the teacher acts not as a distant authority, but as a partner in learning [33].

References

1. Acker, V.: The French Educator Célestin Freinet (1896–1966): An Inquiry into How his Ideas Shaped Education. Lexington Books (2007)
2. Alcock, L., Inglis, M.: Doctoral students' use of examples in evaluating and proving conjectures. Educ. Stud. Math. **69**(2), 111–129 (2008). https://doi.org/10.1007/s10649-008-9149-x
3. Alexander, R.J.: A Dialogic Teaching Companion. Routledge (2020). https://doi.org/10.4324/9781351040143
4. Almeida, D.: Mathematics undergraduates' perceptions of proof. Teach. Math. Appl. Int. J. IMA **14**(4), 171–177 (1995). https://doi.org/10.1093/teamat/14.4.171
5. Andrade, H.G.: Teaching with rubrics: the good, the bad, and the ugly. Coll. Teach. **53**(1), 27–31 (2005). https://doi.org/10.3200/CTCH.53.1.27-31
6. Artigue, M., Blomhøj, M.: Conceptualizing inquiry-based education in mathematics. ZDM **45**(6), 797–810 (2013). https://doi.org/10.1007/s11858-013-0506-6
7. Bachelard, G.: La formation de l'esprit scientifique: contribution à une psychanalyse de la connaissance. J. Vrin, Bibliothèque des textes philosophiques (1993)
8. Bell, T., Witten, I., Fellows, M.: Computer science unplugged: offline activities and games for all ages. Computer Science Unplugged (1998)
9. Black, P., Wiliam, D.: Assessment and classroom learning. Assess. Educ. Principles Policy Pract. **5**(1), 7–74 (1998). https://doi.org/10.1080/0969595980050102
10. Carless, D., Salter, D., Yang, M., Lam, J.: Developing sustainable feedback practices. Stud. High. Educ. **36**(4), 395–407 (2011). https://doi.org/10.1080/03075071003642449
11. Chatain, J., Müller, C., Chatain, K., Calabrese, L., Kapur, M.: Concreteness and abstraction in mathematics education: a taxonomy of the semantic landscape. Educ. Psychol. Rev. **37**(4), 1–49 (2025). https://doi.org/10.1007/s10648-025-10073-9
12. Durand-Guerrier, V.: Which notion of implication is the right one? From logical considerations to a didactic perspective. Educ. Stud. Math. **53**(1), 5–34 (2003). https://doi.org/10.1023/A:1024661004375

13. Durand-Guerrier, V., Boero, P., Douek, N., Epp, S.S., Tanguay, D.: Examining the role of logic in teaching proof. In: Proof and Proving in Mathematics Education: The 19th ICMI Study, pp. 369–389. Springer, Dordrecht (2012). https://doi.org/10.1007/978-94-007-2129-6_16
14. Freeman, S., et al.: Active learning increases student performance in science, engineering, and mathematics. Proc. Natl. Acad. Sci. **111**(23), 8410–8415 (2014). https://doi.org/10.1073/pnas.1319030111
15. Freinet, C.: Education through Work: A Model for Child-Centered Learning, vol. 19. Edwin Mellen Press (1993)
16. Freinet, C.: Les techniques Freinet de l'École Moderne. Bourrelier (1969)
17. Freinet, C.: Œuvres pédagogiques, Seuil, Paris. two-volume edition published under the direction of Madeleine Freinet (1994)
18. Freire, P.: Pedagogy of the oppressed. Herder and Herder (1970), translated by Myra Bergman Ramos
19. Freudenthal, H.: Mathematics as an Educational Task. Springer, Heidelberg (1973). https://doi.org/10.1007/978-94-010-2903-2
20. Goos, M.: Learning mathematics in a classroom community of inquiry. J. Res. Math. Educ. **35**(4), 258–291 (2004). https://doi.org/10.2307/30034910
21. Guzdial, M.: Learner-Centered Design of Computing Education: Research on Computing for Everyone. Morgan & Claypool Publishers (2015). https://doi.org/10.1007/978-3-031-02216-6
22. Harel, I., Papert, S.: Constructionism: Research Reports and Essays, 1985–1990. Ablex Publishing (1991). https://hal.science/hal-00692031
23. Hazzan, O.: How students attempt to reduce abstraction in the learning of mathematics and in the learning of computer science. Comput. Sci. Educ. **13**(2), 95–122 (2003). https://doi.org/10.1076/csed.13.2.95.14202
24. Jacquet-Francillon, F.: Le centenaire de Célestin Freinet. Spirale-Revue de recherches en éducation **24**(1), 286–289 (1999). https://www.persee.fr/doc/spira_0994-3722_1999_num_24_1_1534_t14_0286_0000_2
25. Kinnear, G., Iannone, P., Davies, B.: Student approaches to generating mathematical examples: comparing e-assessment and paper-based tasks. Educ. Stud. Math. **119**(2), 179–201 (2025). https://doi.org/10.1007/s10649-024-10361-1
26. Knuth, D.E.: Computer programming as an art. Commun. ACM **17**(12), 667–673 (1974). https://doi.org/10.1145/361604.361612
27. Kuznetsova, E., Régnier, J.C.: Individualization of educational process according to C. Freinet: a pilot experiment in a group of language learners. Procedia-Soc. Behav. Sci. **154**, 87–91 (2014). https://doi.org/10.1016/j.sbspro.2014.10.117
28. Lavasani, M.G., Khandan, F.: The effect of cooperative learning on mathematics anxiety and help seeking behavior. Procedia. Soc. Behav. Sci. **15**, 271–276 (2011). https://doi.org/10.1016/j.sbspro.2011.03.085
29. Lave, J., Wenger, E.: Situated Learning: Legitimate Peripheral Participation. Cambridge University Press, Cambridge (1991). https://doi.org/10.1017/CBO9780511815355
30. Laxmi, G.: Flipped learning in mathematics classroom: a systematic literature review. Brixton Scholarly Rev. **2**(1), 135–149 (2025). https://doi.org/10.3126/bsr.v2i1.78176
31. Lemke, J.L.: Talking Science: Language, Learning, and Values. Ablex Publishing (1990)
32. Mazur, E., Hilborn, R.: Peer Instruction: A User's Manual. Prentice Hall, Upper Saddle River (1997). https://doi.org/10.1063/1.881735

33. Nowak-Fabrykowski, K.: Freinet's concept of teachers and theory of teaching. McGill J. Educ./Revue des sciences de l'éducation de McGill **27**(001) (1992). https://mje.mcgill.ca/article/view/8015
34. Pan, Y., Harun, J.: Conquering coding fears: a systematic review of programming anxiety in higher education. J. Inf. Technol. Educ. Res. **24**, 020 (2025). https://doi.org/10.28945/5543
35. Papert, S.: Mindstorms: Children, Computers, and Powerful Ideas. Basic Books Inc. (1980). https://doi.org/10.5555/1095592
36. Peyronie, H.: Freinet, 70 ans après: Une pédagogie du travail et de la dédicace? Presses universitaires de, Caen (2017)
37. Rabazas Romero, T., Ramos Zamora, S., Sanz Simón, C.: Freinet pedagogy in the university: an innovative project in the History of Education. Paedagog. Hist. **55**(4), 589–607 (2019). https://doi.org/10.1080/00309230.2018.1556308
38. Shulman, L.S.: Those who understand: knowledge growth in teaching. Educ. Res. **15**(2), 4–14 (1986). https://doi.org/10.3102/0013189X015002004
39. Sierpinska, A.: Understanding in Mathematics. Routledge, Abingdon (1994). https://doi.org/10.4324/9780203454183
40. Smalley, R.T., Hopkins, S.: Social climate and help-seeking avoidance in secondary mathematics classes. Aust. Educ. Research. **47**(3), 445–476 (2020). https://doi.org/10.1007/s13384-020-00383-y
41. Stylianides, A.J.: The notion of proof in the context of elementary school mathematics. Educ. Stud. Math. **65**(1), 1–20 (2007). https://doi.org/10.1007/s10649-006-9038-0
42. Sweller, J.: Cognitive load theory, learning difficulty, and instructional design. Learn. Instr. **4**(4), 295–312 (1994). https://doi.org/10.1016/0959-4752(94)90003-5
43. Tall, D.: Advanced Mathematical Thinking, vol. 11. Springer, Cham (2006). https://doi.org/10.1007/0-306-47203-1
44. Temple, C., de Yunén, A.M.H., Montenegro, L.: Reading around the world: the "Global Method" of Célestin Freinet: whole language in a European setting? Read. Teach. **48**(1), 86–89 (1994). http://www.jstor.org/stable/20201370
45. Topping, K.: Peer assessment between students in colleges and universities. Rev. Educ. Res. **68**(3), 249–276 (1998). https://doi.org/10.3102/00346543068003249
46. Vygotsky, L.S.: Mind in Society: The Development of Higher Psychological Processes. Harvard University Press (1978). https://doi.org/10.2307/j.ctvjf9vz4

Interrelations and Overlaps Between Computational and Algebraic Thinking: A Cross-National Study in Schools

Daranee Lehtonen[1](✉), Esa Satomaa[1], Valentina Dagienė[2], Javier Bilbao[3], Tuğba Öztürk[4], and Marika Parviainen[1,2]

[1] Turku Research Institute for Learning Analytics, University of Turku, Turku, Finland
{daranee.lehtonen,esa.v.satomaa,mhparv}@utu.fi

[2] Institute of Educational Sciences, Vilnius University, Vilnius, Lithuania
valentina.dagiene@mif.vu.lt

[3] University of the Basque Country, Bilbao, Spain
javier.bilbao@ehu.eus

[4] Ankara University, Ankara, Türkiye
tozturk@ankara.edu.tr

Abstract. Computational thinking (CT) and algebraic thinking (AT) are increasingly recognised as key 21st-century competencies. While both are central to education and society, their interrelationship remains underexplored. This mixed-methods study investigated connections between CT and AT using assessment data from 1,694 students aged 9–14 across six countries. Quantitative analyses—including correlation, linear regression, and confirmatory factor analysis (CFA)—showed moderate overall correlations between CT and AT, with associations strengthening with age. A seven-factor CFA model (CT and six AT subskills) provided the best fit across all age groups. Content analysis of assessment items indicated conceptual and procedural overlaps between the two domains. The findings suggest that successful completion of CT tasks may involve cognitive processes commonly associated with AT, and vice versa. Overall, these results enhance understanding of CT–AT relationships and offer practical guidance for curriculum design, assessment, and interdisciplinary pedagogy in schools.

Keywords: Computational Thinking · Algebraic Thinking · Mathematical Thinking · Assessment · Cross-National Study

1 Introduction

The rapid digitalisation of society has reshaped the skills needed for success in education, employment, and civic life. Educational systems are now expected not only to impart subject knowledge, but also to develop higher-order cognitive abilities such as reasoning, problem solving, creativity, and adaptability. Among these key competencies, computational thinking (CT)—the process of formulating problems and solutions in ways that can be effectively executed by humans or computers—has emerged as a crucial skill,

M. Löffler et al. (Eds.): CMSC 2026, LNCS 16951, pp. 257–272, 2027.
https://doi.org/10.1007/978-3-032-38887-2_19

while algebraic thinking (AT)—which involves recognising, representing, and generalising mathematical structures and relationships—has long been central to mathematics education. Both are valued for their methodical approach to problem solving and their role in building analytical proficiency [1].

CT and AT are recognised as essential 21st-century skills. CT enables students to formulate problems and design solutions executable by humans or computers, while AT facilitates the understanding and representation of mathematical structures and relationships [1]. Although grounded in different academic traditions, CT and AT share cognitive characteristics and are increasingly seen as complementary ways of reasoning. With the rise of new technologies, such as data mining and artificial intelligence, understanding the interplay between CT and AT has become a key research area across mathematics, computer science, and the broader STEM disciplines.

Recent educational reforms have foregrounded the integration of CT into school curricula, often alongside mathematics and science [2]. AT, meanwhile, is understood as a developmental process that begins well before formal algebra instruction and underpins students' mathematical development. As educational systems move toward interdisciplinary learning, examining the connections between CT and AT can inform curriculum design, assessment, and instructional practices.

1.1 Computational Thinking

CT is widely understood as a problem-solving approach that involves formulating problems and solutions in ways that can be effectively executed by humans or computers [3]. Under this perspective, CT is considered a foundational competency for all learners, not just those pursuing computer science. It equips individuals with essential skills for navigating and solving complex problems in the digital era, including decomposition, pattern recognition, abstraction, and algorithmic thinking. In recent years, CT has been integrated into curricula worldwide, extending beyond computer science into subjects such as mathematics, science, and engineering [4].

Educational initiatives inspired by the Bebras framework [5] have promoted the development and assessment of CT in schools, with an emphasis on logical reasoning, efficient problem-solving strategies, and the design and evaluation of algorithms. CT evolves with age: younger students typically employ more concrete approaches, while older students develop more abstract and formal reasoning skills [2, 6].

1.2 Algebraic Thinking

AT is widely considered fundamental to mathematical literacy and advanced mathematics. Modern perspectives view AT as extending far beyond algebraic symbol manipulation or equation solving, instead highlighting a broad range of cognitive processes. These include identifying and generalising mathematical structures and relationships; representing generalisations with symbols, drawings, or graphs; and reason about and model with those symbolised generalisation [7, 8].

Lehtonen et al. review identifies several key sub-skills of AT: generalised arithmetic (generalising structures and relationships), equivalence and equations (manipulating equivalent and inequivalent statements), functional thinking (generalising patterns and relationships between co-varying quantities), representation (expressing generalisations using various modalities), transformation (modifying expressions to maintain equivalence), and transversal skills (e.g., problem solving and modelling) [9].

AT development begins in early grades and progresses as students shift from arithmetic reasoning to more abstract thinking [9]. During this period, learners evolve from reasoning about specific examples to articulating general rules and relationships. The age range of 9–14 is especially important, as students consolidate foundational algebraic concepts and begin working with more formal mathematical representations [10].

1.3 Interrelations Between Computational and Algebraic Thinking

Although CT and AT are often treated separately in curricula and research, their potential connections and mutual benefits have attracted increasing interest. Emerging evidence suggests that CT and AT overlap conceptually and procedurally and may share common cognitive processes [11–14]. For example, generalising patterns in algebra aligns with abstraction in CT, and solving equations often involves abstraction, algorithmic thinking, decomposition, and evaluation. Conversely, CT may also strengthen students' mathematical reasoning and problem-solving abilities.

Recent studies further highlight these links. Bråting and Kilhamn identified common cognitive processes—such as generalisation, abstraction, and representation—across CT and AT [15]. Ferreira et al. found significant associations between CT and mathematical reasoning [11], while Öztürk et al. documented increasing research activity connecting CT and mathematical thinking [14]. Additionally, Sarmasági et al. reported positive educational outcomes when CT and AT are integrated in K–12 education [1].

1.4 A Subsection Sample

While interest in the relationship between CT and AT is growing, several important research gaps remain. Most existing studies are conceptual, theoretical, or based on case studies and small-scale interventions, resulting in limited large-scale empirical evidence from diverse student populations. Additionally, prior research often treats CT and AT as broad constructs, confirming their association but offering little insight into how specific sub-skills overlap or relate to student performance. Clarifying links between particular AT and CT skills is essential for developing more nuanced theoretical and educational models. Finally, cross-national studies are rare; most research is confined to single countries, limiting understanding of the generalisability and consistency of findings across different curricula and cultures.

Addressing these gaps is crucial for advancing educational theory and practice, as understanding how CT and AT are related can inform integrated curriculum design, assessment frameworks, and insights into how students transfer reasoning strategies across domains. This study investigates the interrelation and overlaps between CT and AT in students aged 9–14 across six countries through two research questions:

RQ1: Is there a significant interrelation between CT and AT skills among students aged 9–14?

RQ2: Which AT skills are present in CT assessment tasks, and which CT skills are present in AT assessment tasks?

To address these questions, the study combines quantitative methods (correlation analysis, regression modelling, and confirmatory factor analysis) with qualitative content analysis of assessment items. This mixed-methods approach offers a comprehensive understanding of the CT–AT relationship and the underlying cognitive processes.

2 The Present Study

2.1 Settings and Participants

The study was conducted from autumn 2024 to spring 2025 with students aged 9–14 in Finland, Hungary, Lithuania, Spain, Sweden, and Türkiye. Of the 3,130 students who participated, only those who completed both CT and AT assessments and provided gender and age information were included, resulting in a final sample of 1,694 students (see Appendix A). Research ethics were ensured through approval from relevant authorities, informed consent from guardians, and full GDPR compliance.

2.2 COMATH Assessment Instrument

The COMATH assessment instrument, developed for the CT&MathABLE Project (https://www.fsf.vu.lt/ct-math-able#assessment-instruments), evaluates CT and AT in students aged 9–14 using three age-specific versions: COMATH1 (ages 9–10), COMATH2 (ages 11–12), and COMATH3 (ages 13–14). Grounded in educational theories and curriculum standards from participating countries and validated by CT and AT experts [16], COMATH comprises a CT test (e.g., algorithmic thinking, abstraction, decomposition, generalisation) and an AT test (generalised arithmetic; equivalence, equations, and inequations; functional thinking; representation; transformation; and transversal skills). COMATH1 has 12 CT items ($\alpha = .85$) and 51 AT items ($\alpha = .95$); COMATH2, 12 CT items ($\alpha = .86$) and 54 AT items ($\alpha = .96$); and COMATH3, 14 CT items ($\alpha = .89$) and 59 AT items ($\alpha = .97$). Cronbach's alphas are deflation-corrected using Somers' delta [17], recommended for dichotomous data. The instrument was administered on the ViLLE digital platform [18], with 45 min for each test.

2.3 Statistical Analysis

To address RQ1, we conducted statistical analyses using combined pilot data. Participants were divided into three age groups, each completing a distinct version of the COMATH test, with analyses performed separately by group. The maximum score for each test corresponded to the total number of items (one point per item).

Descriptive statistics were calculated for the entire sample as well as for key covariate subgroups (see Appendix B1–B3). The relationship between CT and AT was examined via Pearson correlation coefficients (r) between both tests' total scores. Additionally, partial correlations were computed, controlling for gender, age, and country, by correlating the residuals from regressions predicting CT and AT scores from these covariates. Two multiple linear regression models were used to test (1) if CT total scores and covariates significantly predicted AT total scores and (2) if AT total scores and covariates significantly predicted one predicting CT total scores. Finally, confirmatory factor analysis

(CFA) was conducted, testing both a two-factor model (CT and AT) and a seven-factor model (CT and six AT subtests). Given the binary nature of the items, the Diagonally Weighted Least Squares (DWLS) estimator was employed. Model fit was evaluated using the Comparative Fit Index (CFI), Tucker–Lewis Index (TLI), and Root Mean Square Error of Approximation (RMSEA); scaled fit indices are reported. All analyses were performed using R version 4.4.2.

2.4 Content Analysis

To address RQ2, a content analysis of COMATH test items was conducted to examine the overlap between CT and AT by identifying the cognitive skills and processes embedded in each item. Although COMATH distinguishes CT and AT items, solving tasks in both domains may draw on similar skills. Using relevant literature, a coding framework encompassing CT (e.g., algorithmic thinking) and AT (e.g., generalisation) dimensions was developed (see Appendix C). Each item was independently coded for cognitive skills and processes relevant to successful task completion.

A purposive sample of COMATH3 items was selected for analysis, as statistical results indicated the highest CT–AT correlation in this age group. Some of these tasks were also used to assess other age groups. This approach aimed to maximise the identification of potentially overlapping cognitive processes; however, the resulting qualitative findings are exploratory and may not represent all age groups. From the CT test, one algorithmic thinking item (ALG) and one item assessing other CT skills (OTH) were chosen. From the AT test, six items representing each AT subskill (AT_1–AT_6) were selected. These items were analysed using the coding framework to identify AT skills presenting in CT tasks, and CT skills presenting in AT tasks.

3 Results

3.1 RQ1: Interrelation Between CT and aT Skills of Students Aged 9–14

Correlation and Partial Correlation. As shown in Table 1, CT and AT total scores showed a moderate positive correlation across all age groups, strongest in the oldest group ($r = .63$) and weakest in the youngest ($r = .48$). Partial correlations, controlling for age, gender, and country, were similar to the standard estimates, suggesting these variables had minimal confounding effect on the CT–AT association.

Table 1. Pearson and Partial Correlation Coefficients (r, $r_{ab.c}$) for CT and AT Total Scores.

Age Group	r [95% CI]	$r_{ab.c}$ [95% CI]
1	.48 [.42, .54]	.47 [.41, .53]
2	.57 [.52, .62]	.59 [.53, .64]
3	.63 [.57, .68]	.62 [.56, .67]

Multiple Linear Regression. The analyses indicated a statistically significant association between CT and AT total scores across all age groups after controlling for age, gender, and country (for detailed results, see Appendix D1–D3). In Age Group 1, AT total score predicted CT total score, $B = 0.13$, $SE = 0.01$, $p < .001$, indicating that a one-point increase in AT was associated with an average increase of 0.13 points in CT, holding other variables constant. Conversely, CT total score predicted AT total score, $B = 1.77$, $SE = 0.14$, $p < .001$. In Age Group 2, AT total score predicted CT total score, $B = 0.14$, $SE = 0.01$, $p < .001$, while CT total score predicted AT total score, $B = 2.44$, $SE = 0.13$, $p < .001$. For Age Group 3, regression coefficients were $B = 0.16$, $SE = 0.01$, $p < .001$ for AT predicting CT, and $B = 2.40$, $SE = 0.14$, $p < .001$ for CT predicting AT. Note that regression coefficients should not be directly compared across age groups, as the total scores reflect different numbers of test items for each test version.

Confirmatory Factor Analysis (CFA). CFA supported an association between CT and AT across all age groups. In Age Group 1, the two-factor model (CT and AT) showed a non-optimal fit, chi-squared test statistic $\chi^2(1889) = 3984.6$, $p < .001$, CFI = .835, TLI = .830, RMSEA = .043. The estimated correlation between CT and AT latent factors was moderate ($r = .58$). The seven-factor model (CT and six AT subfactors) demonstrated improved, though still non-optimal fit, $\chi^2(1869) = 3455.5$, $p < .001$, CFI = .875, TLI = .870, RMSEA = .038. Among AT subfactors, correlations with CT ranged from .41 (lowest, for representation) to .59 (highest, for functional thinking), with a mean of .51 (see Table 2 for all subfactors).

Table 2. CT–AT Subfactor Correlations in Seven-Factor CFA Models by Age Group.

	AT_1	AT_2	AT_3	AT_4	AT_5	AT_6	*M*
CT Age Group 1	.46	.54	.59	.41	.49	.55	.51
CT Age Group 2	.61	.63	.71	.54	.69	.67	.64
CT Age Group 3	.66	.67	.69	.76	.67	.70	.69

AT_1 = Generalised Arithmetic; AT_2 = Equivalence, Equations, and Inequations; AT_3 = Functional Thinking; AT_4 = Representation; AT_5 = Transformation; AT_6 = Transversal Skills.

In Age Group 2, the two-factor model exhibited acceptable fit, $\chi^2(2078) = 3956.1$, $p < .001$, CFI = .897, TLI = .894, RMSEA = .038. The CT–AT factor correlation was strong ($r = .69$). The seven-factor model also demonstrated acceptable fit, $\chi^2(2058) = 3738.3$, $p < .001$, CFI = .908, TLI = .904, RMSEA = .036. AT subfactor correlations with CT ranged from .54 (lowest, for representation) to .71 (highest, for functional thinking), with a mean of .64. In Age Group 3, both the two-factor and seven-factor models showed acceptable fit (two-factor: $\chi^2(2554) = 4314.6$, $p < .001$, CFI = .907, TLI = .904, RMSEA = .038; seven-factor: $\chi^2(2534) = 4084.3$, $p < .001$, CFI = .918, TLI = .915, RMSEA = .036). The CT–AT factor correlation was strong ($r = .73$). In this group, representation had the highest correlation with CT ($r = .76$), possibly reflecting the greater number of Representation items in COMATH3 compared to other versions.

Generalised arithmetic had the lowest correlation ($r = .66$). The mean correlation across AT subfactors was .69, with little variation between subfactors. Overall, CT and AT latent factors were moderately to strongly associated across age groups. All AT subfactors showed at least moderate correlations with CT.

All analyses indicated an association between CT and AT, weakest in the youngest age group (moderate) and strongest in the oldest (moderate to strong), with associations persisting after controlling for age, gender, and country. CFA showed that all AT subfactors were at least moderately associated with CT across age groups, with particularly small differences among subfactors in Age Group 3.

3.2 RQ2: Overlap Between CT and aT Skills for Task Completion

Analysis of CT Test Items. Two CT items were selected for content analysis: one assessing algorithmic thinking and one targeting other CT skills. The ALG-10-A item (Fig. 1) asks students to identify switch combinations that produce a specific output in a logic-gate circuit. This assesses algorithmic thinking, along with decomposition and abstraction, as students analyse the circuit step by step and reason about input–output relationships. From an AT perspective, representation supports task completion, as logic gates and switches visually model logical relationships. Functional thinking helps determine how outputs depend on inputs, while generalisation is demonstrated when students apply rules for logic gate behaviour across different circuit sections.

The game "Light on" has 8 switches that can be operated. Wires lead out of these switches, which lead through some components and finally to a light bulb.

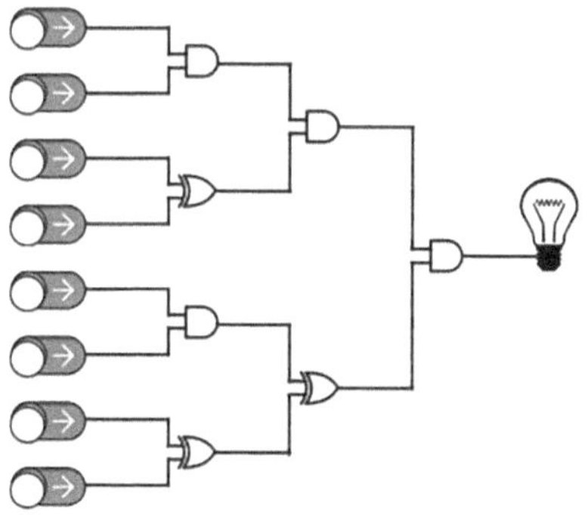

The output from the component is ON only when BOTH incoming wires are ON.

The output from the component is ON when exactly ONE of the incoming wires is ON.

Which switches have to be ON for the light bulb to be on? Click the switches to turn them on.

Fig. 1. ALG-10-A item assessing algorithmic thinking for Age Groups 2 and 3.

The OTH-12-A item (Fig. 2) asks students to identify and apply the rule governing how candies move through a tube. This item primarily assesses abstraction and pattern recognition, as students focus on the underlying structure and reason about input–output relationships. From an AT perspective, representation aids completion, as the candies serve as pictorial variables. The task also involves generalisation and functional thinking, as students must infer the general rule—that the output sequence is the reverse of the input—reflecting core AT.

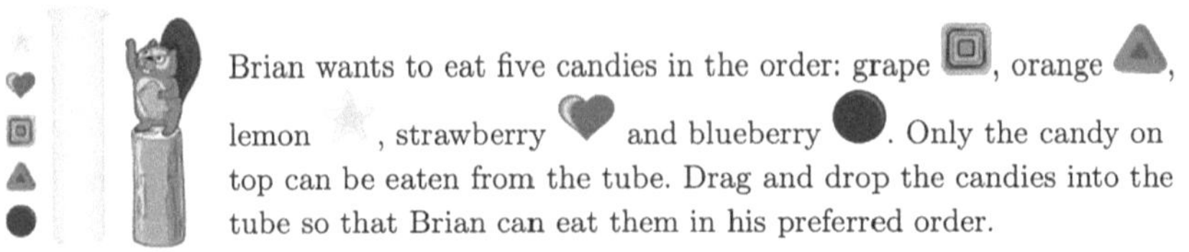

Fig. 2. OTH-12-A item assessing abstraction and pattern recognition for Age Groups 1–3.

Analysis of AT Test Items. Six AT items, one from each subdomain (generalised arithmetic; equivalence, equations, and inequations; functional thinking; representation; transformation; and transversal skills), were selected for content analysis. The AT3_1.3 item (Fig. 3) asks students to identify the mathematical expression representing a turtle's movement on a number grid. This primarily assesses generalisation, as students infer and apply movement rules (e.g., moving down one row increases the value by 10) rather than counting squares individually. From a CT perspective, solving this item requires interpreting the movement as a sequence of actions (algorithmic thinking), breaking it into steps (decomposition), and recognising numerical patterns in the grid (pattern recognition).

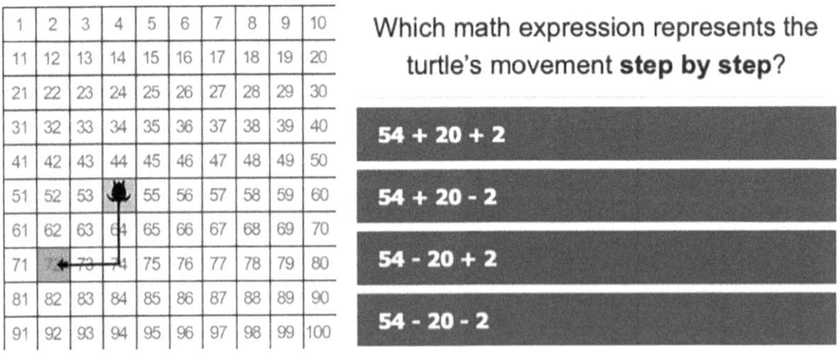

Fig. 3. AT3_1.3 item assessing generalisation of mathematical structure and relationships for Age Groups 2 and 3.

The AT3_2.5 item (Fig. 4) requires students to determine the price of a pencil using the given information. This primarily assesses students' ability to form and solve an equation: $4x + (x + 1) = 11$, where pencil = x, ruler = x + 1. From a CT perspective, the item involves following a sequence of steps to find the unknown (algorithmic thinking), decomposing the problem into sub-tasks (identifying item relationships, calculating totals, solving for x), and abstracting from irrelevant visual details.

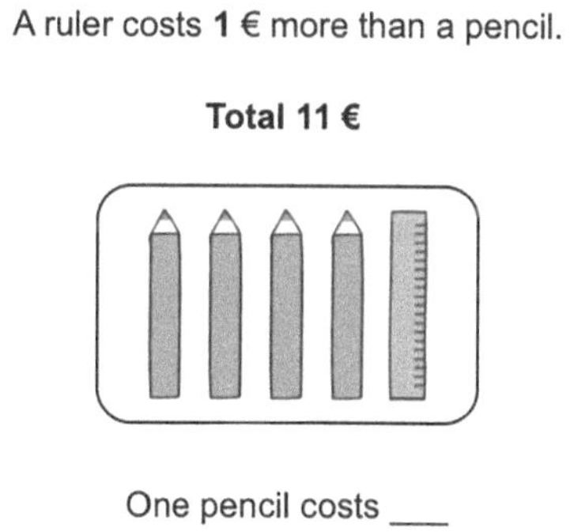

Fig. 4. AT3_2.5 item assessing equation-solving skills for Age Groups 1–3.

The AT3_3.2 item (Fig. 5) primarily assesses functional thinking by asking students to continue an arithmetic sequence. Students must recognise that each increment increases by one (+1, +2, +3,...), so the next operation is +4, yielding the term 5. The item draws on pattern recognition and generalisation from a CT perspective, as students detect regularities, formulate the recursive rule, and extend the sequence.

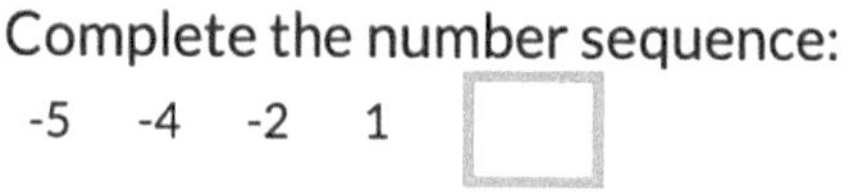

Fig. 5. AT3_3.2 item assessing functional thinking for Age Group 3.

The AT3_4.1 item (Fig. 6) asks students to select the expression for e, with the correct answer being $(f - 4) \div 2$. It assesses representation skills by requiring coordination of a visual model (where f is divided into e, e, and 4) and its symbolic counterpart capturing the structural relationship ($f = e + e + 4$). From a CT perspective, the task involves abstraction, decomposition, and generalisation. Students abstract quantitative relationships from the visual model, decompose f into its parts, and determine the sequence of operations needed to isolate e—subtracting 4 from f and dividing by 2.

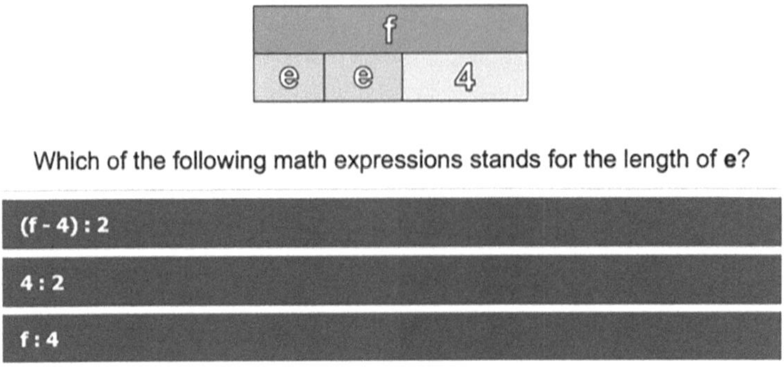

Fig. 6. AT3_4.1 item assessing use of multiple representations for Age Groups 2–3.

The AT3_5.2 item (Fig. 7) assesses transformation skills by asking students to simplify a mathematical expression. This involves identifying like terms and combining them to produce an equivalent, simplified form. From a CT perspective, the task involves pattern recognition to identify repeated variables (f) and cancelling constants (-3 and $+ 3$), abstraction to focus on algebraic relationships rather than the order of terms, and algorithmic thinking to apply algebraic rules step by step.

If **b** is a number, b + b can be simplified (written more simply) as 2b.

Simplify the math expression.

5f - 2f - 3 + f + 3 + g

Fig. 7. AT3_5.2 item assessing skill in transforming expressions for Age Group 3.

The AT3_6.2 item (Fig. 8) assesses transversal skills by presenting an image of a traditional mobile with interconnected rods and asking students to determine the weight of a target figure. This requires understanding balance, analysing quantitative relationships among the components, and inferring unknown weights based on the distribution and equilibrium within the structure. From a CT perspective, the task involves decomposition—breaking the mobile into manageable substructures, analysing weight relationships within each balanced section, and integrating partial solutions. Generalisation is also required, as students must recognise and apply the balancing principle across different levels of the mobile.

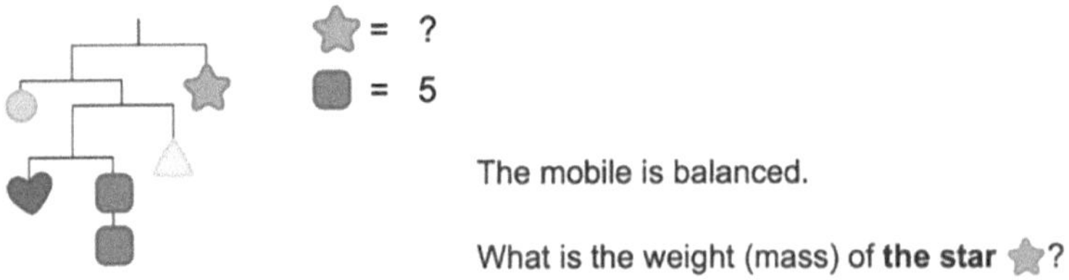

Fig. 8. AT3_6.2 item assessing transversal skills such as reasoning for Age Groups 2–3.

4 Discussion and Conclusions

Findings from this mixed-methods cross-national study reveal that CT and AT are interrelated and share significant conceptual and procedural overlaps. Quantitative analyses show consistent, moderate-to-strong associations across age groups, aligning with prior research (e.g., [11, 12]). These associations persisted even after controlling for age, gender, and country, confirming their robustness across student backgrounds. The relationship was strongest among older students, suggesting stronger connections with increased mathematical and computational experience. CFA showed that all AT subfactors were at least moderately linked with CT, with only minor subfactor differences at higher grade levels, indicating a convergence of skills.

Content analysis, in line with [14, 15], revealed conceptual and procedural overlaps between CT and AT. For example, block-based environments like Scratch engage both CT skills (e.g., algorithmic thinking, decomposition, abstraction) and AT skills (e.g., symbolic representation, functional thinking, variable use). As noted by Ferreira et al. [11] and Wu and Yang [13], both CT and AT rely on logic and abstraction for problem solving, planning, and analysis, indicating that the two domains share several underlying cognitive processes. However, the present correlational design does not allow causal conclusions regarding the development of either domain.

Despite limitations related to the focus on students aged 9–14 and the content analysis of COMATH test items, this study enhances understanding of CT–AT connections and provides empirical support for the benefits of integrating CT and AT in educational practice, particularly for curriculum design, assessment, and interdisciplinary pedagogy. Fostering CT and AT prepares students for academic success and develops transferable 21st-century skills. As Sarmasági et al. [1] suggest, CT can be introduced naturally through mathematical processes such as classification and comparison, while AT can be strengthened through tasks involving abstraction and pattern recognition. Programming languages can also foster AT and related awareness [15].

Given the moderate to strong associations and overlaps between CT and AT, future studies should investigate the potential causal relationships between these domains, as understanding causality could reveal whether strengthening one supports development of the other. Research is also needed to identify effective strategies for integrating CT and AT in schools and to evaluate the impact of such integration on student learning outcomes. Additionally, future work should focus on supporting teachers with targeted professional development and resources to facilitate the effective implementation of integrated approaches.

Acknowledgments. This study is part of the EDUCA Flagship Programme funded by the Research Council of Finland (#358924, #358947) and the Ministry of Education and Culture (Doctoral school pilot #VN/3137/2024-OKM-4). The authors thank CT&MathABLE project partners: Ankara University, Eötvös Loránd University, Gedminų Progymnasium, KTH Royal Institute of Technology, University of the Basque Country, University of Turku, and Özkent Akbilek Middle School. Special thanks to the international Bebras community for their contributions to Bebras task development, which informed the COMATH CT design. Generative AI assisted with language revision; the authors take full responsibility for the content.

Disclosure of Interests. The authors declare no competing interests related to this article.

Appendix

Appendix A. Student Distribution by Age Group, Country, and Total Sample Size

Age Group	Finland	Hungary	Lithuania	Spain	Sweden	Türkiye	Total
1	6	52	415	84	0	34	591
2	54	35	352	100	40	51	632
3	110	84	174	0	71	32	471
Total	170	171	941	184	111	117	1694

Appendix B1. Descriptive Statistics for CT and AT Total Scores: Age Group 1

Group	CT		AT		*n*
	M (*SD*)	Min, Max	*M* (*SD*)	Min, Max	
All students	6.06 (2.51)	0, 12	24.78 (9.40)	3, 49	591
Girls	6.25 (2.60)	0, 12	24.28 (9.23)	3, 47	294
Boys	5.87 (2.42)	0, 12	25.28 (9.55)	4, 49	297
Finland	6.50 (2.43)	4, 10	30.33 (10.42)	11, 41	6
Hungary	5.02 (2.97)	0, 12	23.10 (9.57)	5, 49	52
Lithuania	6.14 (2.42)	0, 12	25.68 (9.46)	3, 47	415
Spain	6.95 (2.36)	1, 11	23.13 (8.29)	10, 47	84
Sweden	–	–	–	–	0
Türkiye	4.35 (2.03)	0, 8	19.41 (8.31)	3, 34	34

Appendix B2. Descriptive Statistics for CT and AT Total Scores: Age Group 2

Group	CT		AT		*n*
	M (*SD*)	Min, Max	*M* (*SD*)	Min, Max	
All students	5.75 (2.57)	0, 12	28.97 (10.43)	6, 54	632
Girls	5.85 (2.37)	1, 11	27.28 (9.96)	8, 51	321
Boys	5.64 (2.76)	0, 12	30.70 (10.63)	6, 54	311
Finland	6.30 (2.83)	0, 12	31.94 (9.01)	7, 49	54
Hungary	6.83 (2.35)	1, 11	30.80 (11.25)	9, 47	35
Lithuania	5.38 (2.51)	0, 12	27.68 (10.42)	6, 53	352
Spain	5.80 (2.24)	1, 11	29.00 (9.50)	12, 50	100
Sweden	7.83 (2.32)	1, 12	32.20 (10.00)	12, 51	40
Türkiye	5.24 (2.55)	1, 11	30.84 (12.05)	9, 54	51

Appendix B3. Descriptive Statistics for CT and AT Total Scores: Age Group 3

Group	CT		AT		*n*
	M (*SD*)	Min, Max	*M* (*SD*)	Min, Max	
All students	7.10 (3.18)	0, 14	31.08 (11.93)	1, 55	471
Girls	7.21 (3.13)	0, 14	30.43 (12.24)	1, 55	238
Boys	6.97 (3.22)	0, 14	31.76 (11.60)	4, 54	233
Finland	6.48 (2.90)	1, 13	31.74 (10.55)	8, 53	110
Hungary	7.38 (2.50)	1, 13	31.37 (10.56)	1, 52	84
Lithuania	7.03 (3.67)	0, 14	30.24 (13.31)	5, 55	174
Spain	–	–	–	–	–
Sweden	8.65 (2.68)	1, 14	32.79 (11.31)	4, 54	71
Türkiye	5.38 (2.23)	1, 10	28.91 (13.16)	7, 49	32

Appendix C. Coding Scheme for Content Analysis

Domain Domain	Dimension	Cognitive Skills and Processes
Computational Thinking (CT)	Algorithmic Thinking	Constructs, applies, or assesses a sequence of steps for problem solving
	Abstraction	Identifies and extracts relevant information while disregarding irrelevant details
	Decomposition	Divides a problem into simpler sub-tasks for easier resolution
	Pattern Recognition	Recognises patterns, similarities, differences, or repetitions in data or problems
Algebraic Thinking (AT)	Generalisation	Recognises and uses general patterns or rules across different contexts
	Functional Thinking	Analyses how changes in one variable affect another and recognises functional relationships
	Representation	Expresses relationships using multiple representations (e.g., symbols, graphs, diagrams)
	Transformation	Modifies representations or structures to preserve their underlying meaning or equivalence
	Abstract Reasoning	Manipulates abstract concepts and relationships independent of concrete representations

Appendix D1. Linear Regression Results for Age Group 1

Predictor Variable	AT predicting CT model			CT predicting AT model		
	B (*SE*)	*t*	*p*	*B* (*SE*)	*t*	*p*
Intercept	2.97 (0.37)	8.05	<0.001***	11.80 (1.37)	8.61	<0.001***
Total Score	0.13 (0.01)	12.97	<0.001***	1.77 (0.14)	12.97	<0.001***
Age:						
> 10 y	0.88 (0.58)	1.52	0.13	1.34 (2.17)	0.62	0.54
10 y	0.33 (0.33)	0.99	0.32	3.97 (1.23)	3.24	0.001**
9 y	0.09 (0.33)	0.28	0.78	1.11 (1.22)	0.91	0.36
Gender (Boy)	−0.54 (0.18)	−3.03	0.003**	1.63 (0.66)	2.46	0.014*
Country:						
Finland	−0.05 (0.88)	−0.05	0.96	4.81 (3.30)	1.46	0.15
Hungary	−0.97 (0.34)	−2.89	0.004**	−0.82 (1.27)	−0.65	0.52
Spain	1.01 (0.27)	3.67	<0.001***	−5.42 (1.01)	−5.35	<0.001***
Türkiye	−1.23 (0.41)	−2.98	0.003**	−4.40 (1.55)	−2.85	0.005**

* $p < .05$. ** $p < .01$. *** $p < .001$.

Appendix D2. Linear Regression Results for Age Group 2

Predictor Variable	AT predicting CT model			CT predicting AT model		
	B (*SE*)	*t*	*p*	*B* (*SE*)	*t*	*p*
Intercept	1.49 (0.35)	4.27	<0.001***	12.21 (1.39)	8.79	<0.001***
Total Score	0.14 (0.01)	18.09	<0.001***	2.44 (0.13)	18.09	<0.001***
Age:						
> 12 y	−0.13 (0.58)	-0.22	0.83	−4.11 (2.41)	−1.70	0.09
12 y	1.00 (0.33)	3.00	0.003**	−1.40 (1.39)	−1.01	0.32
11 y	0.02 (0.31)	0.06	0.95	1.45 (1.28)	1.14	0.25
Gender (Boy)	−0.71 (0.16)	-4.40	<0.001***	4.04 (0.66)	6.11	<0.001***
Country:						
Finland	0.04 (0.31)	0.12	0.91	3.19 (1.29)	2.48	0.014*
Hungary	0.74 (0.36)	2.03	0.04*	0.16 (1.51)	0.11	0.91
Spain	0.49 (0.23)	2.15	0.03*	−0.51 (0.96)	−0.53	0.60
Sweden	1.18 (0.36)	3.32	<0.001***	0.53 (1.49)	0.36	0.72
Türkiye	−0.82 (0.31)	-2.68	0.008**	4.66 (1.27)	3.68	<0.001***

* $p < .05$. ** $p < .01$. *** $p < .001$.

Appendix D3. Linear Regression Results for Age Group 3

Predictor Variable	AT predicting CT model			CT predicting AT model		
	B (*SE*)	*t*	*p*	*B* (*SE*)	*t*	*p*
Intercept	2.00 (0.36)	5.59	<0.001***	12.53 (1.31)	9.53	<0.001***
Total Score	0.16 (0.01)	16.86	<0.001***	2.40 (0.14)	16.86	<0.001***
Age:						
> 14 y	1.06 (0.32)	3.34	<0.001***	2.87 (1.24)	2.32	0.02*
14 y	0.56 (0.25)	2.24	0.03*	−0.99 (0.97)	−1.02	0.31
Gender (Boy)	−0.47 (0.22)	−2.15	0.03*	1.87 (0.85)	2.19	0.03*
Country:						
Finland	−0.83 (0.30)	−2.76	0.006**	1.63 (1.17)	1.39	0.16
Hungary	0.29 (0.31)	0.94	0.35	0.09 (1.22)	0.07	0.94
Sweden	1.42 (0.34)	4.19	<0.001***	−1.17 (1.34)	−0.88	0.38
Türkiye	−1.13 (0.46)	−2.48	0.013*	2.62 (1.78)	1.47	0.14

* $p < .05$. ** $p < .01$. *** $p < .001$.

References

1. Sarmasági, P., et al.: Bridging algebraic and computational thinking: impacts on student development in K–12 education. Inf. Educ. **24**(2), 343–376 (2025)
2. Dagienė, V., Kampylis, P., Giannoutsou, N., Engelhardt, K., Malagoli, C., Bocconi, S.: Fostering computational thinking in compulsory education in Europe: a multiple case study. Baltic J. Mod. Comput. **12**(2), 189–221 (2024)
3. Wing, J.M.: Computational thinking. Commun. ACM **49**(3), 33–35 (2006)
4. Whitney-Smith, R. M.: The emergence of computational thinking in national mathematics curricula: an Australian example. J. Pedagogical Res. **7**(2) (2023)
5. Dagienė, V.: Solving Bebras-like tasks: approaches for concept building. In: Fernau, H., Schwank, I., Staub, J. (eds.) Creative Mathematical Sciences Communication (CMSC) 2024. Lecture Notes in Computer Science, vol. 15229. Springer, Cham (2024)
6. Jasutė, E., Parviainene, M., Dagienė, V.: Computational thinking in technology-supported primary mathematics: a systematic review. Baltic J. Mod. Comput. **14**(2), 328–355 (2026)
7. Kaput, J. J.: What is algebra? What is algebraic reasoning? In: Kaput, J., Carraher, D.W., Blanton, M.L. (eds.), Algebra in the Early Grades, pp. 5–17. Routledge (2008)
8. Kieran, C.: Algebraic thinking in the early grades: what is it? Math. Educ. **8**(1), 139–151 (2004)
9. Lehtonen, D., Laakso, M.-J., & Räsänen, P.: Assessing algebraic thinking in K–9 education: a scoping review of assessment contexts, instruments, and competence domains [Manuscript submitted for publication] (2026)
10. Broley, L., Buteau, C., Modeste, S., Rafalska, M., Stephens, M.: Computational thinking and mathematics. In: Pepin, B., Gueudet, G., Choppin, J. (eds.), Handbook of Digital Resources in Mathematics Education, pp. 1–38. Springer, Heidelberg (2023)

11. Ferreira, T.D., dos Santos, J.S., de Araújo Medeiros, R., Andrade, W.L., Brunet, J., Melo, M.R.A.: Exploring the relationship of mathematical reasoning and computational thinking. In: Workshop de Informática na Escola (WIE), pp. 583–591 (2023)
12. Qolfathiriyus Firdaus, A., Suryanti, S., Hadi, F.N.: Identification of computational thinking patterns in students' mathematical problem solving: empirical evidence from algebra learning activities. Jurnal Riset Pendidikan Dan Inovasi Pembelajaran Matematika **10**(1), 22–32 (2026)
13. Wu, W.-R., Yang, K.-L.: The relationships between computational and mathematical thinking: a review study on tasks. Cogent Educ. **9**(1), 1–19 (2022)
14. Öztürk, T., Güven, İ., Ayantaş, T.: Bibliometric review of computational thinking and mathematical thinking. In: The Annual Conference of Association for Learning Technology (ALTC) 2024, pp. 1–7 (2024)
15. Bråting, K., Kilhamn, C.: Exploring the intersection of algebraic and computational thinking. Math. Think. Learn. **23**(2), 170–185 (2021)
16. Dagienė, V., Lehtonen, D., Satomaa, E., Laakso, M.J.: COMATH: development and validation of a cross-national assessment instrument for computational thinking in primary and secondary education. Inf. Educ. **25**(1), 59–108 (2026)
17. Somers, R.H.: A new asymmetric measure of association for ordinal variables. Am. Sociol. Rev. **27**(6), 799–811 (1962)
18. Laakso, M.-J., Kaila, E., Rajala, T.: ViLLE – Collaborative education tool: designing and utilizing an exercise-based learning environment. Educ. Inf. Technol. **23**, 1655–1676 (2018)

From Problem Solving to Research Thinking: A Graph-Theoretic Whodunit

Lucas Lorieau[1(✉)] and Nicolas Schivre[2]

[1] CNRS - LIMOS, Université Clermont Auvergne, Clermont-Ferrand, France
lucas.lorieau@limos.fr
[2] LIMOS, Université Clermont Auvergne, Clermont-Ferrand, Clermont-Ferrand, France
nicolas.schivre@doctorant.uca.fr

Abstract. Popularizing research thinking methods is critical in a society where skepticism towards science has grown throughout the years. Due to the dramatic spread of generative AI systems, the need for such popularization in Computer Science (CS) has become more and more important. We propose an activity aiming at demystifying the methodology used by CS researchers through a whodunit puzzle based on the novel "Who killed the Duke of Densmore?" by Claude Berge. After introducing the problem to participants, along with the definition of interval graphs and some of their basic properties, we invite the participants to model the problem using graphs. The culprit's identity is to be revealed through a series of simple reasoning on interval graphs. The whole activity displays the main steps of research thinking in CS, namely concept definition, property analysis, modeling and mathematical reasoning. We discuss in detail the activity and its realization in outreach events for middle- and high-schoolers.

Keywords: CS Unplugged · Interval Graphs · Scientific Outreach · Research Thinking

1 Introduction

Computer science unplugged (CS Unplugged) has been, in recent years, a growing part of Computer Science related scientific outreach. It was introduced by Bell, Witten and Fellows [2], and follows the principle of introducing concepts of Computer Science without using computers or programming languages. It typically aims at popularizing a wide range of concepts occurring for example in cybersecurity, algorithmics or even machine learning. Yet, a significantly smaller portion of CS Unplugged work has been dedicated to so-called research thinking. Research thinking can be defined as the set of classical mental processes and

This work was supported by the French "Agence Nationale de la Recherche" project SPAWN (ANR-25-CE48-1750) and has received financial support from the CNRS through the MITI interdisciplinary programs.

M. Löffler et al. (Eds.): CMSC 2026, LNCS 16951, pp. 273–283, 2027.
https://doi.org/10.1007/978-3-032-38887-2_20

methodology used in order to solve problems and further develop the knowledge in a specific research field. See [1] for a broader review of CS Unplugged and related popularization projects.

At a time when scientific expertise is questioned in public debates, helping students understand how scientific knowledge is constructed, *i.e.* through scientific thinking, has become a major educational challenge. Scientific knowledge increasingly shapes contemporary societies, and understanding how such knowledge is produced is arguably as important as learning its outcomes. Yet students in middle- and high-school rarely encounter the processes that characterize scientific work: exploring unfamiliar situations, formulating conjectures, identifying relevant evidence, revising ideas, and building convincing arguments. Scientific results often appear as finished products rather than as the outcome of a collective process of inquiry. Providing students with opportunities to experience these practices can help demystify scientific research. Computer Science, through carefully designed problems, offers a particularly fruitful setting in which students can engage in forms of reasoning closely related to those used by researchers. In particular, graph theory has been used to train and teach, through research-based methods, similar thought processes [5].

Graph theory allows to model a variety of problems and networks. Graph algorithms, in particular, have been designed for a range of everyday applications, such as shortest path computation, scheduling, or network monitoring. Several activities of CS Unplugged explore the popularization of Graph Theory through so called "pen-and-paper" activities[1], where the audience is introduced to some algorithms or some definitions on graphs with exclusively on-paper exercises or puzzles.

In this work, we propose a pen-and-paper scientific outreach activity designed for teenagers (in middle-school or high-school) whose goal is to popularize the thinking process of researchers in Graph Theory. To do so, we take inspiration from a novel published in 1994 by Claude Berge named *Who Killed the Duke of Densmore?* [3]. In this novel, a duke was assassinated and the detectives only manage to obtain from the suspects partial information about the murder. Suspects only disclosed who they had seen on the crime scene on the day the murder was committed. Fortunately, this information is sufficient to find out which suspect is guilty, thanks to properties of interval graphs. We adapt the previous story to get an activity where, after presenting the background story of a murder mystery, we define the mathematical objects that will be used to solve the mystery, namely interval graphs and their segment-based representation. A small section of exercises is proposed so that the participants can manipulate the definitions and understand better how both the graph-based and the segment-based representation are equivalent. Finally, the murder is solved through identifying forbidden structures on a graph modeling the information gathered by the

[1] See https://classic.csunplugged.org/activities/ including activites around search in graphs, the Dominating Set and Graph Colouring problems. See also https://sancy.iut.uca.fr/ iso/activites/equivalencegraphes/ where the concept of equivalence of problems on graphs is popularized (in French).

detectives. The conclusion of the activity highlights the steps of reasoning of the activity, and bring them in parallel with the research thinking performed by graph theory researchers.

Note that the graph used to solve the murder is exactly the one present in the original story by Berge, and has been kept as-is for this work. We designed, as an additional material, the exercises present in the Appendix, and integrated the story into a popularization of research thinking.

Some work has been dedicated to the popularization of research processes to a younger audience using discrete mathematics problems. Ouvrier-Buffet presents in [10] an activity based on polyominoes to introduce the concept of mathematical proofs and to have elementary pupils to experiment a simple research process. Modeling and proofs were also popularized in anterior work using discrete mathematics, and in particular graph theory [7].

The paper is organized as follows: Sect. 2 presents some scientific context around interval graphs and states the main results used during the activity, as well as related work concerning CS Unplugged outreach. Then, a thorough description of the activity is proposed in Sect. 3. Finally, we report in Sect. 4 how the activity was presented to teenagers during a national scientific outreach day in France. An Appendix containing a worksheet that was handed out to the participants is also included, for completeness.

2 Preliminaries

In this section, we present some graph notations and some properties of the interval graphs that will be used throughout the activity. Please note that understanding these notions does not require any specific technical background. A *graph* is a pair $G = (V, E)$ where $V = \{v_1, v_2, \ldots, v_n\}$ is a set of n *vertices* and $E \subseteq V \times V$ a set of m *edges*. An *interval graph* is a graph which can be represented as a *set of intervals* on the *real line*. More specifically, each interval corresponds to a vertex and an edge appears when two intervals intersect on the real line. In conclusion, interval graphs have two representations: a *graph-based* and a *segment-based* representation. See Fig. 1 for examples of interval graphs and one of their segment-based representation.

Note that, for a single interval graph, many segment-based representations are valid.

Interval graphs are a broadly studied graph class due to its theoretical and real life applications. Several characterizations of interval graphs can be found in the literature [8]. Recognizing whether a given graph is an interval graph is tractable [4], and many problems that are computationally hard in general are tractable on interval graphs [6,9]. Applications using interval graphs range from scheduling problems, as intervals can easily represent time, to computational biology and network overseeing [9].

The key property of interval graphs that will be of main interest for the activity is the fact that some graphs do not admit a segment-based representation, *i.e.* they are not interval graphs. The simplest example of such a graph is the

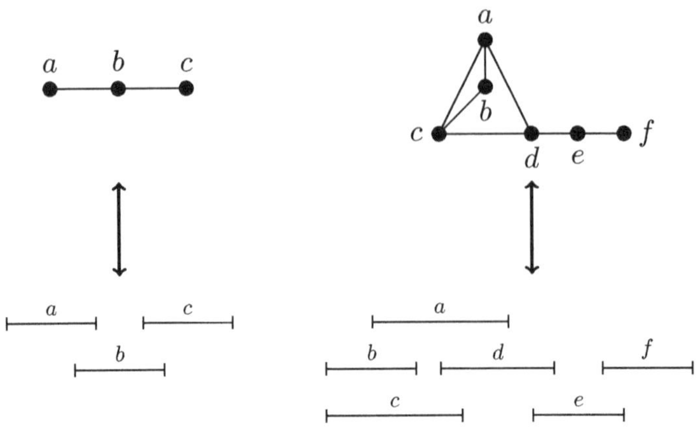

Fig. 1. Examples of interval graphs and one of their segment-based representation.

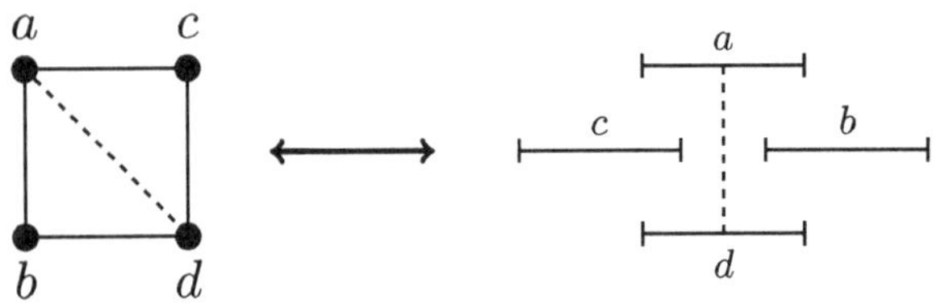

Fig. 2. The C_4 does not admit any segment-based representation and hence is not an interval graph. The dashed edge on the left part is required for the left segment-based representation to be valid.

(induced) cycle on four vertices, also denoted by C_4. Figure 2 illustrates the fact that no segment-based representation can represent a C_4.

Finally, we describe a simple property that will be used (and explained informally) during the activity in order to solve the murder case. For a set S of vertices of a graph $G = (V, E)$, we say that the subgraph of G *induced* by S, and denote by $G[S]$, the graph on vertices of S where the edges of $G[S]$ are exactly the edges of G whose both endpoints are in S. We say that a graph H is an *induced subgraph* of $G = (V, E)$ if there exists a set $S \subseteq V$ such that $H = G[S]$. It is easy to see that if a graph G is an interval graph, removing a vertex of G along with its adjacent edges will yield an interval graph. Thus, if some induced subgraph of G cannot be represented using segments, then G is not an interval graph. In particular, if a graph contains C_4 as an induced subgraph, then it is not an interval graph.

We also give another forbidden structure of interval graphs. The *neighborhood* a vertex $v \in V$ in graph $G = (V, E)$ is the set of vertices that share an edge of E with v. A *path* in G is a sequence of vertices of V such that any two neighboring vertices of the sequence share an edge of E. An *Asteroidal Triple* (AT) is a set S of three vertices such that there exists a path between each pair of S that avoids the neighborhood of the third vertex of S. Such a structure is known to be forbidden in interval graphs [8].

3 Activity Description

In this section, we describe the main features of the activity and precise some important steps of the presentation.

Introduction. For the remainder, we suppose that the participants have no prior knowledge in graph theory. The first step of the activity aims at explaining the context of the murder story. At this step, we do not disclose any information about the testimonies that will be used in order to solve the murder. We only present the context and state the apparent main goal of the activity: finding the murderer. We will explicit links with research in Graph Theory at the end of the activity, as a conclusion. We also state at this point that we will introduce the audience to a new mathematical tool, namely graphs, in order to solve the murder.

Graphs and Properties. To introduce the concept of graph to the participants, we stress the need for some *modeling* of the situation, in order to easily express all the information gathered during the investigation. At this stage, we define the notion of graphs. Immediately, we exemplify this notion to the participants by using analogies with real-life. More specifically, we explain how graphs can model concepts or objects (such as human, house,...) with vertices, as well as relations between those concepts or objects (such as friendship, reachability,...) with edges. Afterwards, we introduce the concept of interval graph and their segment-based representation and have the participants practice their understanding of definitions with some exercises. They are required to find a segment-based representation of an interval graph and draw an interval graph based on its segment-based representation. Those exercises can be found in the Appendix. Finally, we briefly highlight that there exists some graphs that are not interval graph, taking C_4 as an example (see Fig. 2) that we introduce as a *forbidden structure*. We try to draw a segment-based representation of C_4 one interval at a time, and convince the participants that no representation using segments is possible as we try to draw correctly the fourth segment.

Murder Case Description We recall the context and give more details to the participants about the murder case. To do so, we clearly state the problem faced (note that modifying the context of the story in order to adapt it to the public targeted is recommended in order to further encourage active participation of your audience): someone has been found dead and the police has been able to identify 8 suspects. The detectives were also able to get some information from each of the suspects: they all claim that they came only once during the murder day on the crime scene, and each suspect gave the list of the other suspects they saw on the crime scene. Moreover, due to the way the victim died, we know that

Table 1. Table of information used to construct the main graph of the activity.

	Alice	Bob	Carole	Donald	Eve	Farid	Gontrand	Héloïse
Alice		x	x		x	x	x	
Bob	x		x					x
Carole	x	x		x	x			x
Donald			x		x			
Eve	x		x	x		x		
Farid	x				x			
Gontrand	x							x
Héloïse		x	x				x	

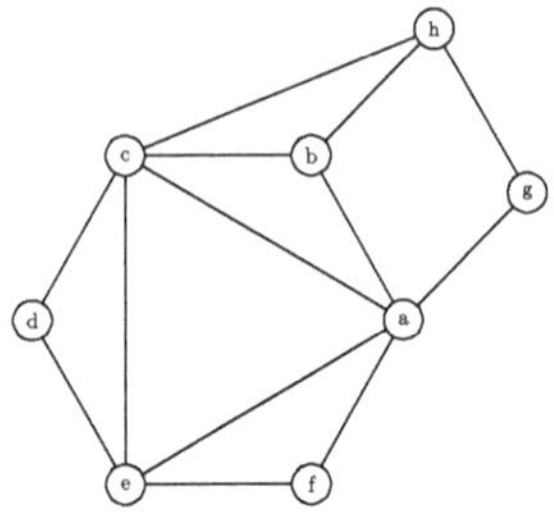

Fig. 3. The graph corresponding to the testimonies.

the murderer must have come at least twice at the place to prepare the crime. For an immersive story, feel free to imagine why the murderer came at least twice (in the case of our story, the murderer came three times) such as hiding a bomb, prepare a bait for the victim, and finally detonate it.

Testimonies and Modeling We then give the precise testimony of each of the suspects to the public, and have them compile and summarize it into a table (see Table 1). Once the table is filled, we check quickly, together with all the participants, that no two testimonies are contradictory by checking whether the table is symmetric. We then highlight that if each suspect only came once on the crime scene, their stay on the crime scene can be drawn an interval from the beginning and the end of their stay on some timeline. It means that we should be able to draw an interval graph based on the testimonies such that each suspect is represented by a unique interval. With this information in mind, we have the participants draw the graph whose adjacency matrix is the table filled previously. Each suspect is a vertex and there is an edge between two suspects if they have seen each other on the crime scene. The resulting graph is represented in Fig. 3.

Solve the Case. Now, we let the audience find by themselves the forbidden structures of the graph, and recall that we presented in the first part of the presentation one forbidden structure. This allows the participants to find two induced C_4

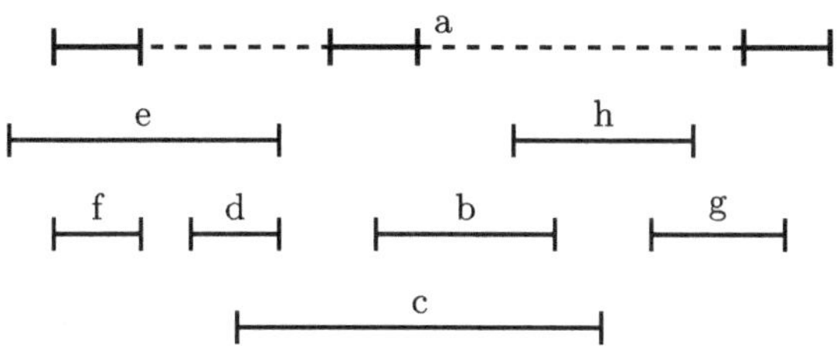

Fig. 4. A segment-based representation of the murder timeline. Alice had to come, in fact, three times for the testimonies to be coherent.

in the graph, namely $[a, b, h, g]$ and $[a, c, h, g]$. Each forbidden structure found in the graph is some piece of evidence reducing the number of suspects. Indeed, if we find a forbidden structure, we know that one of the suspects associated to vertices of the forbidden structure must have lied. After finding both induced C_4 of the graph, the set of possible liars is restricted to three ($[a, h, g]$), but no other induced C_4 is present in the graph. A new structure has to be identified, and we give the participants, after a moment, the structure to be found and sketch a proof that it cannot be represented with segments, which is enough to designate the culprit which is in fact Alice. In fact, this structure is $G[\{a, b, c, d, e, f\}]$. This subgraph is a forbidden structure because it contains an AT. Indeed, it can be verified that the set $\{b, d, f\}$ is an AT. For the sake of simplicity, we do not introduce this notion to the participants, and we try to draw, as previously done for the C_4, a valid interval representation of $G[\{a, b, c, d, e, f\}]$. First, we notice that $[a, c, e]$ are all connected together, and draw then three lines of the same length each above the others. Then, to add b, we extend a and c a bit to the left such that b does not cross e. We do the same for d to the right with c and e. Finally, we ask the participant how they would add f from now on, and, most of the time, a participant says that it is not possible. It remains to use this new evidence to reduce the amount of suspects. Indeed, only Alice is in the three identified forbidden structures, and is thus guilty. Depending on the time left for the activity, it is possible but not mandatory to show how the actual, valid interval representation of the situation. In reality, Alice came three times to the crime scene (see Fig. 4).

Conclusion. We end the activity by giving a brief summary of what the participants have learned, and what we could achieve using the concept of interval graph. As a final word, we explicit further how to map the steps of the activity with research.

- We started from a specific problem and look for some model that could represent it, here graphs.
- We defined some concepts and prove results or properties on those. Here, we defined the class of interval graphs and showed that some graphs were not a part of this class. We also stress at this point that this class of graphs is used in real life applications, as stated in the introduction of this paper.

- Then, using the model associated to the original problem, we performed some reasoning in order to obtain some information on the problem, and eventually solve it. Here, we spotted some forbidden structures that helped reducing the list of suspects.

We end the session by explaining that this thought process is central in scientific thinking and that the participants, for the time of the presentation, were put in a "guided" research situation and applied very similar thought process to the ones used by researchers.

4 Activity Report

Every year, a national scientific outreach event called *La Fête de la Science* ("The Science Festival") is organized in France in order to promote science and encourage scientists to share their knowledge with the general public. We presented the activity described in this paper to middle-schoolers and high-schoolers during this event in Clermont-Ferrand, France. Around ten groups of about 15 students participated throughout the day, and we spent with each group around 45 min. Those students have *a priori* no knowledge of graph theory before the activity, and usually do not have a substantial prior experience with algorithms. The students were accompanied by teachers (usually not mathematics teachers) and we had them also participating in the activity. The activity was well received by a majority of the participants. Thanks to the early introduction of the challenge of the activity (namely finding the culprit), the students were curious and eager to solve the mystery. No particular misconception by the participants were noticed, but for a few groups, the link between a forbidden structure and how to reduce the number of potential culprits had to be explained twice in order for the audience to fully understand this link.

The small exercise session after introducing the different notions in the middle of the activity seemed to help the students to understand the presented notions, since each group quite easily found the first two forbidden structures in the graph of suspects. One hint that had to be given to almost all groups during the exercise session was that edges do not need to be drawn as straight lines, as this (intentionally) never was stated when presenting graphs. This was another way to put participants in a situation where they had to question definitions and assumptions, as a researcher could. All groups could solve the different exercises in the time limit we imposed (around 10 min).

Only one group struggled to find a second C_4. Interestingly, some groups also found without major hints that the C4 structure was not sufficient in order to discriminate between the two last suspects. These groups proposed spontaneously some other structures (or at least expressed that there was no more induced C_4 in the graph), and for one of the groups, the proposition was in fact the correct one.

The majority of the teachers mentioned that the presentation was clear and helpful to demystify research in graph theory and more broadly in Theoretical Computer Science. They specifically enjoyed the progression of the activity, especially the conclusion part that they did not anticipated, as the real goal of the activity was not disclosed before the conclusion.

5 Conclusion

We presented an outreach activity that illustrates, through solving a toy murder case in a CS Unplugged fashion, some core elements of scientific thinking process in Theoretical Computer Science. Further work might naturally try to extend the set of activities aiming at further popularizing research thinking to middle- and high-schoolers.

Acknowledgment. We thank the anonymous reviewers for proposing additional relevant references and providing some valuable feedback on the content of the activity report and the presentation of the activity.

We thank the members of the "*Informatique Sans Ordinateurs*" group of IREM Clermont-Ferrand for their helpful feedback and discussions.

A Exercise Sheet

The material on the next two pages was handed out to the participants in order to have them interact with the notions presented. It contains three small exercises on interval graphs and their representations, as well as some table to help gathering the pieces of information of the investigation.

Draw the corresponding interval graphs

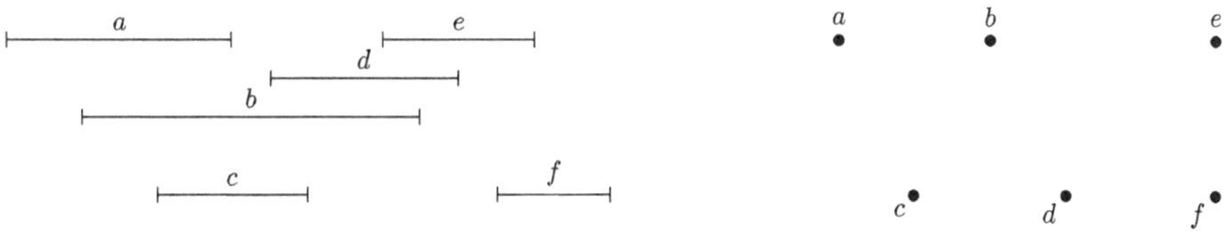

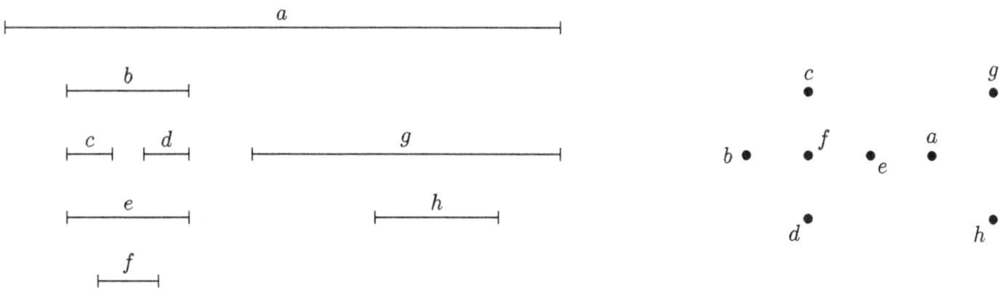

Draw a segment-based representation corresponding to this graph

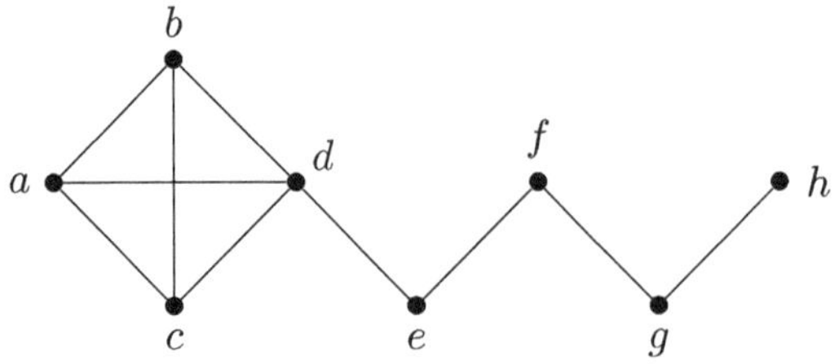

Let's solve the mystery!

	Alice	Bob	Carole	Donald	Eve	Farid	Gontrand	Héloïse
Alice								
Bob								
Carole								
Donald								
Eve								
Farid								
Gontrand								
Héloïse								

References

1. Bell, T., Rosamond, F., Casey, N.: Computer science unplugged and related projects in math and computer science popularization. In: Bodlaender, H.L., Downey, R., Fomin, F.V., Marx, D. (eds.) The Multivariate Algorithmic Revolution and Beyond. LNCS, vol. 7370, pp. 398–456. Springer, Heidelberg (2012). https://doi.org/10.1007/978-3-642-30891-8_18
2. Bell, T., Witten, I., Fellows, M.: Computer Science Unplugged - an enrichment and extension programme for primary-aged children. Computer Science Unplugged (2002)
3. Berge, C.: Qui a tué le duc de Densmore ? No. 67 in Bibliothèque oulipienne, Oulipo, Paris (1994)
4. Cao, Y.: Recognizing (unit) interval graphs by zigzag graph searches. In: Symposium on Simplicity in Algorithms (SOSA), pp. 92–106 (2021)
5. Chinn, P.Z.: Discovery-method teaching in graph theory. In: Gimbel, J., Kennedy, J.W., Quintas, L.V. (eds.) Annals Discrete Mathematics, pp. 375–384. Elsevier (1993). Quo Vadis, Graph Theory?
6. Golumbic, M.C.: Interval graphs. In: Algorithmic Graph Theory and Perfect Graphs, pp. 171–202. Academic Press (1980)
7. Grenier, D., Payan, C.: Discrete mathematics in relation to learning and teaching proof and modelling (1999)
8. Lekkeikerker, C., Boland, J.: Representation of a finite graph by a set of intervals on the real line. Fundam. Math. **51**, 45–64 (1962)
9. McKee, T.A., McMorris, F.R.: Interval graphs. In: Topics in Intersection Graph Theory, pp. 45–64. SIAM (1999)
10. Ouvrier-Buffet, C.: Exploring proof in elementary schools within Research-Situations for the Classroom -the case of polyominoes (2021). https://hal.science/hal-03499721

Author Index

M. Löffler et al. (Eds.): CMSC 2026, LNCS 16951, p. 285, 2027.
https://doi.org/10.1007/978-3-032-38887-2

Kolibri 360 GmbH
Ferdinand-Jühlke-Straße 7
99095 Erfurt, Deutschland
produktsicherheit@kolibri360.de